Study Guide for

Introduction to

HUMAN ANATOMY AND PHYSIOLOGY

Copyright © 1992 W. B. SAUNDERS COMPANY All rights reserved

Study Guide for
Introduction to

HUMAN ANATOMY
AND PHYSIOLOGY

Eldra Pearl Solomon, Ph.D.

Mical K. Solomon, L.M.T.

W.B. SAUNDERS COMPANY
A Harcourt Health Sciences Company
Philadelphia   London   New York   St. Louis   Sydney   Toronto

Copyright © 1992 W. B. SAUNDERS COMPANY All rights reserved

W. B. Saunders Company
A Harcourt Health Sciences Company
The Curtis Center
Independence Square West
Philadelphia, PA 19106-3399

Editor: Michael J. Brown
Designer: Bill Donnelly
Cover Designer: Paul Fry
Production Manager: Peter Faber
Manuscript Editor: Allison Esposito and Amy Norwitz
Illustration Coordinator: Peg Shaw

Study Guide for Introduction to Human Anatomy and Physiology   ISBN 0-7216-3967-4

Copyright © 1992 by W. B. Saunders Company.

All rights reserved. No part of this publication may be reproduced or transmitted in any form or by any means, electronic or mechanical, including photocopy, recording, or any information storage and retrieval system, without permission in writing from the publisher.

Printed in the United States of America.

Last digit is the print number: 9   8

Copyright © 1992 W. B. SAUNDERS COMPANY All rights reserved

Using the Study Guide

Learning anatomy and physiology has been compared to learning a new language. To understand how the body is constructed and how it functions, you must learn the language used by health professionals. The exercises included in *Study Guide for Introduction to Human Anatomy and Physiology* are designed to help you learn both the words and the concepts presented in the textbook, *Introduction to Human Anatomy and Physiology*, by Eldra Pearl Solomon, Ph.D. Working through the exercises, diagrams, and crossword puzzles, you will test your mastery of the material and gain both knowledge and confidence.

Follow the following steps to help you use this study guide effectively.

1. Read the Outline presented at the beginning of the chapter. The Outline reflects the organization of the corresponding textbook chapter and provides you with an overview of the material in the *Study Guide* chapter.

2. Read the Learning Objectives and refer back to them frequently as you work through the chapter exercises. The Learning Objectives tell you what you need to do to demonstrate mastery of the material.

3. Answer the Study Questions provided for each section. After you complete a section, check your answers in the Answer Key at the back of the book. If one or more of your answers are incorrect, reread the corresponding sections in the textbook.

4. Label the diagrams that are provided in appropriate sections throughout the study guide. To check the accuracy of your labels, consult the corresponding labeled diagram presented in the textbook.

5. When you feel confident that you have learned the material in the chapter you are studying, check your level of mastery by completing the Post Test at the end of the chapter.

6. To build your confidence even more, complete the Crossword Puzzle provided for the block of chapters you are studying. Answers are provided in the Answer Key.

Copyright © 1992 W. B. SAUNDERS COMPANY All rights reserved

Contents

One
INTRODUCING THE HUMAN BODY **1**

Two
CELLS AND TISSUES **12**

Three
THE SKIN **23**

Four
THE SKELETAL SYSTEM **33**

Five
THE MUSCULAR SYSTEM **52**

Six
THE CENTRAL NERVOUS SYSTEM **63**

Seven
THE PERIPHERAL NERVOUS SYSTEM **73**

Eight
THE SENSE ORGANS **79**

Nine
ENDOCRINE CONTROL **88**

Ten
THE CIRCULATORY SYSTEM: BLOOD **101**

Eleven
THE CIRCULATORY SYSTEM: THE HEART **109**

Twelve
CIRCULATION OF BLOOD AND LYMPH **118**

Copyright © 1992 W. B. SAUNDERS COMPANY All rights reserved

Thirteen

THE BODY'S DEFENSE MECHANISMS **130**

Fourteen

THE RESPIRATORY SYSTEM **137**

Fifteen

THE DIGESTIVE SYSTEM **147**

Sixteen

THE URINARY SYSTEM **162**

Seventeen

REGULATION OF FLUIDS AND ELECTROLYTES **169**

Eighteen

REPRODUCTION **178**

Answer Key **191**

Copyright © 1992 W. B. SAUNDERS COMPANY All rights reserved

One

INTRODUCING THE HUMAN BODY

OUTLINE

I. The body has several levels of organization
II. The body systems work together to maintain life
III. Metabolism is essential for digestion, growth, and repair of the body
IV. Homeostatic mechanisms maintain an appropriate internal environment
V. The body has a basic plan
    A. Directions in the body are relative
    B. The body has three main planes
    C. We can identify specific body regions
    D. There are two main body cavities
    E. It is important to view the body as a whole

LEARNING OBJECTIVES

After you have studied this chapter, you should be able to:
1. Define anatomy and physiology.
2. List in sequence the levels of biological organization in the human body, starting with the simplest (the chemical level — atoms and molecules) and ending with the most complex (the organism).
3. Describe the 10 principal organ systems.
4. Define metabolism, and contrast anabolism with catabolism.
5. Define homeostasis, describe it as a basic mechanism of human physiology, and give examples.
6. Describe the anatomical position of the human body.
7. Define and use appropriately the principal orientational terms employed in human anatomy.
8. Recognize sagittal, transverse, and frontal sections of the body and of body structures.
9. Define and locate the principal regions and cavities of the body.

Copyright © 1992 W. B. SAUNDERS COMPANY All rights reserved

NAME _____

STUDY QUESTIONS

Within each category, fill in the blanks with the correct response.

I. INTRODUCTION TO ANATOMY AND PHYSIOLOGY

Adapted	Physiology	Stomach
Anatomy	Shape	Structure

1. _____ is the science of body structure.

2. _____ is the science of body function.

3. Each structure of the body is marvelously _____ for carrying out its specific function.

4. The muscular walls of the _____ are especially constructed for churning and breaking down food.

5. In the body, the size, _____, and _____ of each part are related to the job the part must perform.

II. LEVELS OF ORGANIZATION IN THE BODY

Atoms	Ion	Nucleus
Body	Matter	Organelles
Cells	Microscope	Organism
Connective	Molecules	Organs
Functions	Muscle	Tissue
Epithelial	Nervous	Water

1. The simplest level of organization in the body is the chemical level, consisting of _____ and molecules.

2. Atoms are the basic units of _____ .

3. An electrically charged atom is called a(n) _____ .

4. Atoms combine chemically to form _____ .

5. Two atoms of hydrogen combine with one atom of oxygen to form one molecule of _____ .

6. In living organisms, atoms and molecules associate in specific ways to form _____ , the building blocks of the body.

7. Although cells vary in size and shape according to their function, most are so small that they can be seen only with a(n) _____ .

8. Each cell consists of specialized cell parts called _____ .

9. The _____ serves as the information and control center of the cell.

Copyright © 1992 W. B. SAUNDERS COMPANY All rights reserved

10. The next highest level of organization after the cellular level is the
    _____ level.

11. A tissue is a group of closely associated cells specialized to perform particular
    _____ .

12. The four main types of tissue in the body are _____ tissue,
    _____ tissue, _____ tissue, and _____
    tissue.

13. Various types of tissue are organized into _____ .

14. A group of tissues and organs that work together to perform specific functions makes
    up a(n) _____ system.

15. Working together with great precision and complexity, the body systems make up
    the living _____ .

III. LEVELS OF BIOLOGICAL ORGANIZATION IN THE BODY

Anabolism   Building   Nutrients
ATP   Catabolism   Oxygen
Body   Cellular respiration   Synthetic
Breaking down   Metabolism

1. Each body system contributes to the dynamic, carefully balanced state of the
   _____ .

2. The chemical processes that take place within the body are collectively called its
   _____ .

3. Two phases of metabolism are _____ and _____ .

4. Catabolism is the _____ phase of metabolism.

5. Cells obtain energy from food molecules by a complex series of catabolic chemical
   reactions called _____ .

6. Nutrients are slowly broken down, and the energy released is packaged within a
   special energy storage compound called _____ .

7. Cellular respiration requires both _____ and _____ .

8. Anabolism is the _____ or _____ phase of metabolism.

IV. HOMEOSTATIC MECHANISMS

Feedback system   Negative feedback   Regulated
Homeostasis   Positive   Stressor
Negative   Positive feedback

1. Metabolic activities are continuously occurring in every living cell, and they must be
   carefully _____ to maintain a consistent internal environment for the
   body.

Copyright © 1992 W. B. SAUNDERS COMPANY All rights reserved

2. The automatic tendency to maintain a relatively constant internal environment is called _____ .

3. A _____ is a stimulus that disrupts homeostasis, causing stress within the body.

4. A _____ consists of a cycle of events in which information about a change is fed back into the system so that the regulator can control the process.

5. In a _____ system, the response counteracts the inappropriate change, thereby restoring the steady state.

6. Most homeostatic mechanisms in the body are _____ feedback systems.

7. In a _____ system, the variation from the steady state sets off a series of events that intensify the changes.

8. The delivery of a baby is an example of a _____ feedback system.

V. THE BASIC PLAN OF THE BODY

Anatomical
Anterior
Axis
Bilateral symmetry
Caudad
Cephalic
Closer
Cranium

Deep
Distal
Dorsal
Inferior
Lateral
Medial
Midline
Mirror

Posterior
Proximal
Superficial
Superior
Ventral
Vertebral column

1. The body consists of right and left halves that are _____ images; that is, it exhibits _____ .

2. Two other important features of the body are the _____ , or brain case, and the backbone, or _____ .

3. In the _____ position, the body is erect, the eyes are looking forward, the arms are at the sides, and the palms and toes are directed forward.

4. The north pole of the human body is the top of the head, its most _____ point.

5. The south pole of the body is represented by the soles of the feet, its most _____ part.

6. The heart is superior to the stomach because it is _____ to the head.

7. The terms _____ and "craniad" are sometimes used instead of the word "superior."

8. The term _____ is sometimes used instead of the word "inferior."

9. The front (belly) surface of the body is _____ or _____ .

10. The back surface of the body is _____ or _____ .

11. The body _____ is an imaginary line extending from the center of the top of the head to the groin.

Copyright © 1992 W. B. SAUNDERS COMPANY All rights reserved

12. The main superior-inferior body axis is _____, going right through the midline of the body.

13. A structure is medial if it is closer to the _____ of the body than to another structure.

14. A structure is _____ if it is toward one side of the body.

15. When a structure is closer to the body midline or point of attachment to the trunk, it is described as _____ .

16. _____ means farther from the midline or point of attachment to the trunk.

17. Structures located toward the surface of the body are _____ .

18. Structures located farther inward are _____ .

VI. THE THREE MAIN PLANES OF THE BODY

Body planes   Midsagittal plane   Transverse plane
Frontal plane   Sagittal plane

1. Imaginary flat surfaces that divide the body into parts are called _____ .

2. A _____ divides the body into right and left parts.

3. A _____ passes through the body axis and divides the body into two mirror-image halves.

4. A _____ divides the body into superior and inferior parts.

5. A _____ divides the body into anterior and posterior parts.

VII. REGIONS OF THE BODY AND BODY CAVITIES

Abdominal cavity   Diaphragm   Spinal canal
Abdominopelvic   Dorsal   Thoracic
Appendicular   Mediastinum   Torso
Axial   Pelvic cavity   Ventral
Body cavities   Pericardial   Viscera
Cranial cavity   Pleural sacs

1. The body may be subdivided into a(n) _____ portion, consisting of the head, neck, and trunk.

2. The _____ portion of the body consists of the limbs.

3. The _____ consists of the thorax, abdomen, and pelvis.

4. The spaces within the body are called _____ .

5. Body cavities contain the internal organs, or _____ .

6. The two principal body cavities are the _____ cavity and the _____ cavity.

Copyright © 1992 W. B. SAUNDERS COMPANY All rights reserved

7. The dorsal cavity is subdivided into the _____, which holds the brain, and the vertebral, or _____, which contains the spinal cord.

8. The ventral cavity is subdivided into the _____, or chest cavity, and the _____ cavity.

9. The thoracic and abdominopelvic cavities are separated by a broad muscle, the _____, which forms the floor of the thoracic cavity.

10. Divisions of the thoracic cavity are the _____, each containing one lung, and the _____ between them.

11. The heart is surrounded by the _____ cavity.

12. The upper portion of the abdominopelvic cavity is the _____.

13. The lower portion of the abdominopelvic cavity is the _____.

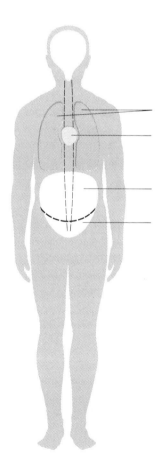

FIGURE 1–1 Fill in the correct labels.

Copyright © 1992 W. B. SAUNDERS COMPANY All rights reserved

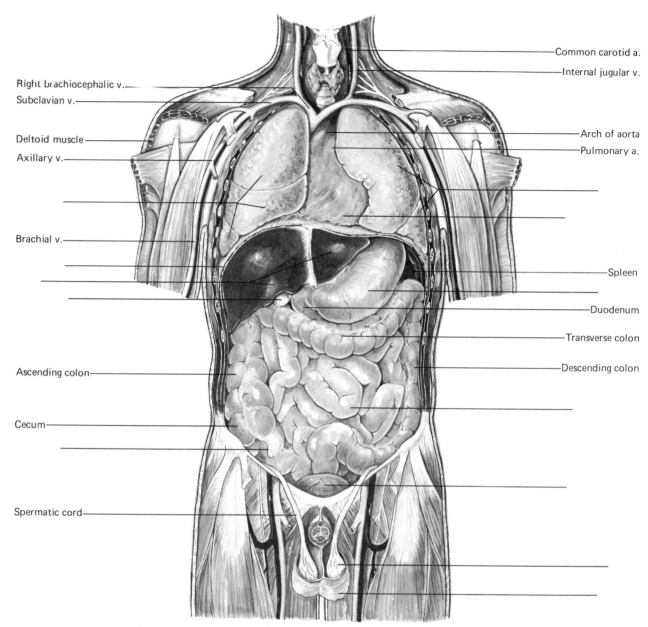

Common carotid a.

Internal jugular v.

Right brachiocephalic v.

Subclavian v.

Deltoid muscle

Axillary v.

Arch of aorta

Pulmonary a.

Brachial v.

Spleen

Duodenum

Transverse colon

Ascending colon

Descending colon

Cecum

Spermatic cord

FIGURE 1–2 Fill in the correct labels.

Copyright © 1992 W. B. SAUNDERS COMPANY All rights reserved

16. The dorsal cavity is subdivided into the _____ cavity, which holds the brain,

    and the _____ cavity, which contains the spinal cord.
    a. Ventral and cranial
    b. Vertebral and spinal
    c. Cranial and vertebral
    d. Ventral and cranial
    e. None of the preceding

17. The ventral cavity is subdivided into the _____ cavity and the _____ cavity.
    a. Thoracic and chest
    b. Thoracic and abdominopelvic
    c. Thoracic and diaphragm
    d. Dorsal and ventral
    e. None of the preceding

Copyright © 1992 W. B. SAUNDERS COMPANY All rights reserved

Two

CELLS AND TISSUES

OUTLINE

   I. The cell contains specialized organelles
  II. Materials move through the plasma membrane by both passive and active processes
 III. Cells divide, forming genetically identical cells
 IV. Tissues are the fabric of the body
     A. Epithelial tissue protects the body
     B. Connective tissue joins body structures
     C. Muscle tissue is specialized to contract
     D. Nervous tissue controls muscles and glands
  V. Membranes cover or line body surfaces

LEARNING OBJECTIVES

After you have studied this chapter, you should be able to:
1. Describe the general characteristics of cells.
2. Describe, locate, and list the functions of the principal organelles and label them on a diagram.
3. Explain how materials pass through cell membranes, distinguishing between passive and active processes.
4. Predict whether cells will swell or shrink under various osmotic conditions.

Copyright © 1992 W. B. SAUNDERS COMPANY All rights reserved

NAME _____

STUDY QUESTIONS

Within each category, fill in the blanks with the correct response.

I. THE CELL CONTAINS SPECIALIZED ORGANELLES

Amino acids	Functions	Plasma membrane
Building blocks	Golgi complex	Proteins
Cellular respiration	Lysosomes	Regulates
Chromosomes	Mitochondria	Ribosomes
Cilia	Nucleolus	Rough
Cytoplasm	Nucleus	Smooth
Endoplasmic reticulum	Organelles	
Enzymes	Ovum	

1. Cells are the living _____ of the body.

2. The size and shape of a cell are related to the specific _____ it must perform.

3. The _____ is one of the largest cells in the human body.

4. The jellylike material of the cell is called _____ .

5. _____ found in the cytoplasm are used to manufacture larger molecules such as proteins.

6. Scattered throughout the cell are specialized _____ that perform different functions within the cell.

7. Every cell is surrounded by a thin membrane called the _____ , which is also called the cell membrane.

8. The plasma membrane protects the cell and _____ the passage of materials into and out of the cell.

9. The _____ is a system of membranes that extends throughout the cytoplasm of many cells.

10. There are two types of endoplasmic reticulum: _____ and

_____ .

11. Rough endoplasmic reticulum has a granular appearance that results from the

presence along its outer walls of organelles called _____ .

12. Ribosomes function as factories in which _____ are manufactured.

13. The _____ is composed of layers of platelike membranes that give it the appearance of a stack of pancakes.

14. An important function of the Golgi complex is to produce _____ .

15. Lysosomes contain powerful digestive _____ that destroy bacteria or other foreign matter.

16. Cells contain tiny power plants called _____ .

Copyright © 1992 W. B. SAUNDERS COMPANY All rights reserved

17. Inside the mitochondria, fuel molecules are broken down, and energy is released; this process is called _____ .

18. _____ are tiny hairlike organelles that project from the surfaces of many types of cells and help move materials outside the cell.

19. The _____ is the control center of the cell.

20. When a cell prepares to divide, the chromatin in the nucleus becomes more tightly coiled and condenses to form rod-shaped bodies called _____ .

21. The _____ is a specialized region within the nucleus where ribosomes are assembled.

II. MATERIALS MOVE THROUGH THE PLASMA MEMBRANE

Active transport   Filtration   Permeable
Diffusion   Osmosis   Phagocytosis

1. The plasma membrane is selectively _____ .

2. _____ is the net movement of molecules or ions from a region of higher concentration to a region of lower concentration brought about by the energy of the molecules.

3. _____ is the diffusion of water molecules through a selectively permeable membrane from a region in which water molecules are more concentrated to a region in which they are less concentrated.

4. _____ is the passage of materials through membranes by mechanical pressure.

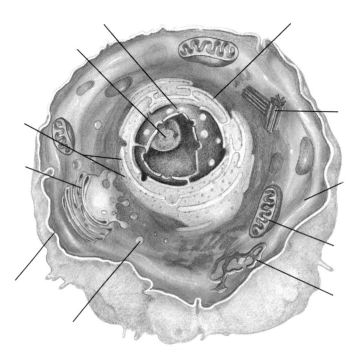

FIGURE 2–1 Fill in correct labels.

Copyright © 1992 W. B. SAUNDERS COMPANY All rights reserved

5. _____ requires cellular energy. In this process, the cell moves materials from a region of lower concentration to a region of higher concentration.

6. In _____ , the cell ingests large, solid particles such as food or bacteria.

III. CELLS DIVIDE, FORMING GENETICALLY IDENTICAL CELLS

Anaphase	Interphase	Prophase
Chromosomes	Metaphase	Telophase
Five	Mitosis	Two

1. Before a cell divides to form two cells, the chromosomes are precisely duplicated, and the cell undergoes _____ .

2. In mitosis, a complete set of _____ is distributed to each end of the parent cell.

3. The life cycle of the cell may be divided into _____ phases.

4. The cell spends most of its life in _____ .

5. _____ is the first stage of mitosis.

6. _____ is the second stage of mitosis.

7. During _____ , the doubled chromosomes separate and begin moving away from one another.

8. _____ begins with the arrival of a complete set of chromosomes at each end of the cell.

9. During telophase, the cell divides, forming _____ cells.

IV. TISSUES ARE THE FABRIC OF THE BODY

Axon	Exocrine	Nervous
Cell body	Gland	Neurons
Connective	Glial	Organ
Contract	Intercellular substance	Protection
Dendrites	Involuntary	Simple
Endocrine	Join together	Smooth
Epithelial	Muscle	Stratified

1. The cells of a tissue produce nonliving materials called _____ , which lie between the cells.

2. _____ tissue protects the body by covering all of its free surfaces and lining its cavities.

3. _____ tissue supports and protects the organs of the body.

4. _____ tissue receives and transmits messages so that the various parts of the body can communicate with one another.

5. _____ tissue is specialized for moving the body and its parts.

6. Epithelial tissue has many functions; its main function is _____ .

Copyright © 1992 W. B. SAUNDERS COMPANY All rights reserved

7. Epithelial tissue may be _____, which is composed of one layer of cells, or _____, which is composed of two or more layers.

8. A(n) _____ consists of one or more epithelial cells that produce and discharge a particular product.

9. Two main types of glands are ____ _____ and _____ glands.

10. The main function of connective tissue is to _____ thc other tissues of the body.

11. Almost every _____ in the body has a supporting framework of connective tissue.

12. Muscle tissue is composed of cells specialized to _____.

13. Cardiac muscle, found in the walls of the heart, is considered _____ because we do not generally make a conscious decision to contract it.

14. _____ muscle is found in the walls of the digestive tract, uterus, blood vessels, and other internal organs. Its fibers are not striated, and its control is involuntary.

15. Nervous tissue consists of _____, cells that are specialized for transmitting nerve impulses, and _____ cells that support and nourish the neurons.

16. A typical neuron has a large _____, which contains the nucleus and from which two types of extensions project.

17. _____ are specialized for receiving impulses, whereas the single _____ conducts information away from the cell body.

V. MEMBRANES COVER OR LINE BODY SURFACES

Membranes	Parietal	Synovial
Mucous	Serous	Visceral

1. _____ are sheets of tissue that cover or line body surfaces.

2. The _____ membrane is a connective tissue membrane that lines the joint cavities.

3. A _____ membrane, or mucosa, lines body cavities that open to the outside of the body.

4. A _____ membrane, or serosa, lines a body cavity that does not open to the outside of the body.

5. The part of the membrane that is attached to the wall of the cavity is the _____ membrane.

6. The part of the membrane that covers the organs inside the cavity is the _____ membrane.

Copyright © 1992 W. B. SAUNDERS COMPANY All rights reserved

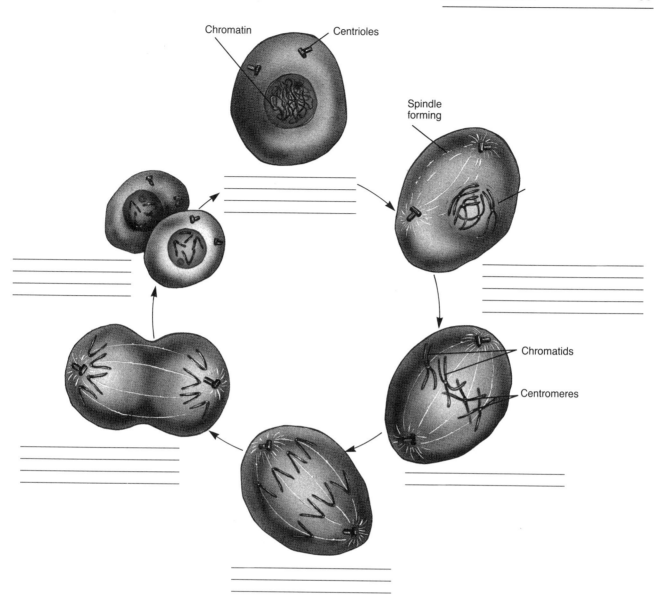

Chromatin   Centrioles

Spindle
forming

Chromatids

Centromeres

FIGURE 2–2 Fill in correct labels.

Copyright © 1992 W. B. SAUNDERS COMPANY All rights reserved

Three

THE SKIN

OUTLINE

I. The skin functions as a protective barrier
II. The skin consists of epidermis and dermis
    A. The epidermis continuously replaces itself
    B. The dermis provides strength and elasticity
    C. The subcutaneous layer attaches the skin to underlying tissues
III. Sweat glands and sebaceous glands are found in the skin
IV. Hair and nails are appendages of the skin
V. Melanin helps determine skin color

LEARNING OBJECTIVES

After you have studied this chapter, you should be able to:
1. List six functions of the skin, and explain how each is important in homeostasis.
2. Compare the structure and function of the epidermis with those of the dermis.
3. Describe the subcutaneous layer.
4. Describe the functions of sweat glands and sebaceous glands.
5. Describe the functions of hair and nails, and describe the structure of a hair.
6. Explain the function of melanin.

Copyright © 1992 W. B. SAUNDERS COMPANY All rights reserved

NAME _____

STUDY QUESTIONS

Within each category, fill in the blanks with the correct response.

I. THE SKIN FUNCTIONS AS A PROTECTIVE BARRIER

Fluid   Sensory receptors   Vitamin D
Integumentary system Sweat

1. Together with the skin glands, hair, and nails, the skin makes up the

   _____ .

2. _____ glands also excrete excess water and some wastes from the body.

3. The skin prevents loss of _____ and thereby prevents cells from drying out.

4. The skin contains a compound that is converted to _____ when it is exposed to the ultraviolet rays of the sun.

5. Located within the skin are _____ that detect touch, pressure, heat, cold, and pain.

II. THE SKIN CONSISTS OF THE EPIDERMIS AND DERMIS

Adipose   Epidermis   Keratin
Collagen   Epithelial   Outer
Connective   Fingerprints   Subcutaneous
Deepest   Glands   Superficial fascia
Dermis   Hair follicles   Upper

1. The outer layer of the skin is called the _____ .

2. The inner layer of the skin is called the _____ .

3. Beneath the skin is an underlying _____ layer.

4. The epidermis consists of stratified _____ tissue.

5. The _____ cells of the epidermis continuously wear off.

6. New epidermal cells are constantly produced in the _____ sublayer of the epidermis.

7. _____ , a tough, waterproofing protein, fills most of each epidermal cell.

8. Dermis consists of dense _____ tissue composed mainly of collagen fibers.

9. _____ is largely responsible for the mechanical strength of the skin.

10. Specialized skin structures such as _____ and _____ are found in the dermis.

11. The _____ portion of the dermis has many small fingerlike elevations that project into the epidermal tissue.

Copyright © 1992 W. B. SAUNDERS COMPANY All rights reserved

12. _____ serve as friction ridges that help us hold on to the objects we grasp.

13. The subcutaneous layer beneath the dermis is also called the _____.

14. Fat stored within the _____ tissue can be mobilized and used as an energy source when adequate food is not available.

III. SWEAT GLANDS AND SEBACEOUS GLANDS ARE FOUND IN THE SKIN

Acne
Body temperature
Ducts
Evaporation

Heat
Pimple
Raise
Sebaceous glands

Sebum
Sweat gland
Water

1. Each _____ is a tiny coiled tube found in the dermis or subcutaneous tissue.

2. Approximately 3 million sweat glands in the skin help maintain _____.

3. Muscle movement and metabolic activity generate _____ and therefore _____ body temperature.

4. Because heat is required for _____, the body becomes cooler as sweat evaporates from the skin.

5. About 1 qt of _____ is excreted in sweat each day.

6. _____, also called oil glands, are usually attached to hair follicles.

7. Sebaceous glands are connected to each hair follicle by little _____ through which they release their secretions.

8. Sebaceous glands secrete an oily substance called _____.

9. At puberty, the stepped-up activity of sebaceous glands can sometimes lead to _____, a condition that is very common during adolescence.

10. Sometimes, the duct of a sebaceous gland ruptures, allowing sebum to spill into the dermis. The skin may become inflamed, and a _____ may form.

IV. HAIR AND NAILS ARE APPENDAGES OF THE SKIN

Capillaries
Contract
Dead
Follicle
Keratin

Nails
Palms
Pink
Protective

Root
Shaft
Smooth
Soles

1. Hair serves a _____ function.

2. Hair is found on all skin surfaces except the _____ and the _____.

Copyright © 1992 W. B. SAUNDERS COMPANY All rights reserved

3. The part of the hair that is visible is called the _____ .

4. The part of the hair that is below the skin surface is called the _____ .

5. The root together with its epithelial and connective tissue coverings are collectively called the hair _____ .

6. At the bottom of the follicle is a little mound of connective tissue containing _____ that deliver nourishment to the cells of the follicle.

7. Each hair consists of cells that manufacture _____ .

8. The shaft of the hair consists of _____ cells and their products.

9. Tiny bundles of _____ muscle are associated with hair follicles.

10. Arrector pili muscles _____ in response to cold or fear, causing the hairs to stand up straight.

11. _____ help protect the ends of the fingers and the toes.

12. Nails appear _____ because of underlying capillaries.

V. MELANIN HELPS DETERMINE SKIN COLOR

Absorbs   Darker   Sun
Albino   Lowest   Sunburned
Cancer   Melanin   Ultraviolet
Color

1. Scattered throughout the _____ layer of the epidermis are cells that produce pigment granules.

2. Pigment granules are composed of a type of protein called _____ .

3. Melanin gives _____ to hair as well as to skin.

4. A(n) _____ is a person of any race who has inherited the inability to produce pigment.

5. Melanin is an important protective screen against the _____ .

6. Melanin _____ harmful ultraviolet rays.

7. Exposure to the sun stimulates an increase in the amount of melanin produced and causes the skin to become _____ .

8. A dark tan is actually a sign that the skin has been exposed to too much _____ radiation.

9. When the melanin is not able to absorb all of the ultraviolet rays, the skin becomes inflamed, or _____ .

10. Over a period of years, excessive exposure to the sun can cause wrinkling of the skin and, sometimes, skin _____ .

Copyright © 1992 W. B. SAUNDERS COMPANY All rights reserved

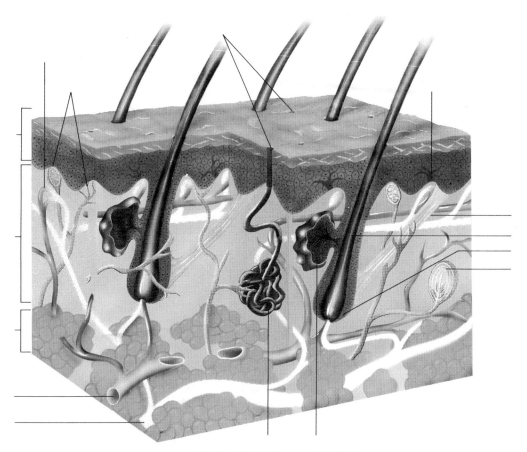

FIGURE 3 – 1 Fill in the correct labels.

Copyright © 1992 W. B. SAUNDERS COMPANY All rights reserved

Four

THE SKELETAL SYSTEM

OUTLINE

    I. The functions of the skeletal system include support and protection
   II. A typical long bone consists of a shaft with flared ends
  III. Bone develops by replacing existing connective tissue
  IV. The skeleton may be divided into the axial and the appendicular skeletons
   V. The skull is the bony framework of the head
  VI. The vertebral column supports the body
 VII. The thoracic cage protects the organs of the chest
VIII. The pectoral girdle attaches the upper limbs to the axial skeleton
  IX. The upper limb consists of 30 bones
   X. The pelvic girdle supports the lower limbs
  XI. The lower limb consists of 30 bones
 XII. Joints are junctions between bones
     A. Joints can be classified according to the degree of movement they permit
     B. A diarthrosis is surrounded by a joint capsule

LEARNING OBJECTIVES

After you have studied this chapter, you should be able to:
1. List five functions of the skeletal system.
2. Label a diagram of a long bone, and describe the microscopic structure of a bone.
3. Contrast endochondral with intramembranous bone development, and describe the role of osteoblasts and osteoclasts in bone production.
4. List and describe the bones of the axial skeleton, and identify each bone on a diagram or skeleton.
5. List and describe the bones of the appendicular skeleton, and identify each bone on a diagram or skeleton.
6. Compare the main types of joints, and describe the structure and functions of a diarthrosis.

Copyright © 1992 W. B. SAUNDERS COMPANY All rights reserved

NAME _____

STUDY QUESTIONS

Within each category, fill in the blanks with the correct response.

I. FUNCTIONS OF THE SKELETAL SYSTEM

Bones   Marrow   Supports
Interaction   Protects   Tendons
Ligaments

1. Two important functions of the skeletal system are that it _____ the body by serving as a bony framework for the other tissues and organs and it _____ delicate vital organs.

2. _____ serve as levers that transmit muscular forces.

3. Muscles are attached to bones by bands of connective tissue called _____.

4. Bones are held together at the joints by bands of connective tissue called

   _____ .

5. The _____ within some bones produces blood cells.

6. The _____ of bones and muscles also makes breathing possible.

II. THE LONG BONE

Bone   Epiphyseal   Metaphysis
Diaphysis   Epiphysis   Periosteum
Endosteum   Hyaline   Yellow

1. The main shaft of a long bone is called its _____ .

2. The _____ is the expanded end of a long bone.

3. In children, a disc of cartilage called the _____ is found between the epiphysis and the diaphysis.

4. The metaphyses are growth centers that disappear at maturity, becoming vague

   _____ lines.

5. Within the long bone, there is a central marrow cavity filled with a fatty connective

   tissue called _____ bone marrow.

6. The marrow cavity of the long bone is lined with a thin layer of cells called the

   _____ .

7. Each long bone is covered by the _____ , a layer of specialized connective tissue.

8. The inner layer of the periosteum contains cells that produce _____ .

9. At its joint surfaces, the outer layer of a bone consists of a thin layer of

   _____ cartilage, the articular cartilage.

Copyright © 1992 W. B. SAUNDERS COMPANY All rights reserved

III. TYPES OF BONE TISSUE

Bone marrow	Epiphyses	Osteons
Canaliculi	Haversian canals	Spindle
Compact	Lacunae	Spongy
Dense	Osteocytes	

1. Two types of bone tissue are _____ bone and _____ bone.

2. Compact bone, which is very _____ and hard, is found near the surfaces of the bone, where great strength is needed.

3. Compact bone consists of interlocking, _____-shaped units called _____, or haversian systems.

4. Within an osteon, _____, the mature bone cells are found in small cavities called _____.

5. Lacunae are arranged in concentric circles around central _____.

6. Threadlike extensions of the cytoplasm of the osteocytes extend through narrow channels called _____. These cellular extensions connect the osteocytes.

7. Spongy bone is found within the _____ and makes up the inner part of the wall of the diaphysis.

8. The spaces within the spongy bone are filled with _____.

IV. BONE DEVELOPMENT AND DIVISION OF THE SKELETON

Apatite	Fetal	Osteoblasts
Appendicular	Finished	Osteoclasts
Axial	Intramembranous	Osteocytes
Bones	Lacunae	Resorption
Endochrondral	Marrow	Tissue
Enzymes	Ossification	Two

1. Bone formation is called _____.

2. During _____ development, bones form in two ways.

3. The long bones develop from cartilage models, a process called _____ bone development.

4. The flat bones of the skull develop from a noncartilage connective tissue scaffold; this is called _____ bone development.

5. _____ are cells that produce bone.

6. A complex calcium phosphate called _____ is present in tissue fluid.

7. As the bone matrix forms around the osteoblasts, they become isolated within small spaces called _____.

8. When osteoblasts become embedded in the bone matrix, they are called _____.

Copyright © 1992 W. B. SAUNDERS COMPANY All rights reserved

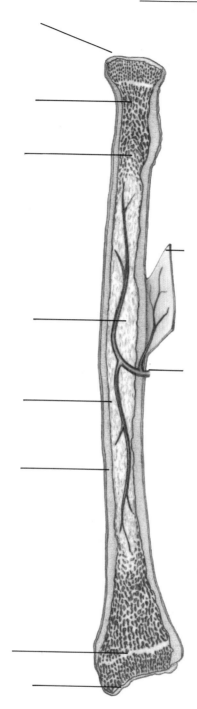

FIGURE 4–1 Fill in the correct labels.

9. As muscles develop in response to physical activity, the _____ to which they are attached thicken and become stronger.

10. As bones grow, bone _____ must be removed from the interior, especially from the walls of the _____ cavity. This process keeps bones from becoming too heavy.

11. _____ are the cells that break down bone, a process called bone

_____ .

Copyright © 1992 W. B. SAUNDERS COMPANY All rights reserved

12. Osteoclasts are very large cells that secrete _____, which digest bone.

13. Osteoclasts and osteoblasts work side by side to shape bones and form the precise grain needed in the _____ bone.

14. The human skeleton can be divided into _____ groups of bones.

15. The _____ skeleton consists of the skull, vertebral column, ribs, and sternum.

16. The _____ skeleton consists of the upper and lower limbs, shoulder girdle, and pelvic girdle (with the exception of the sacrum).

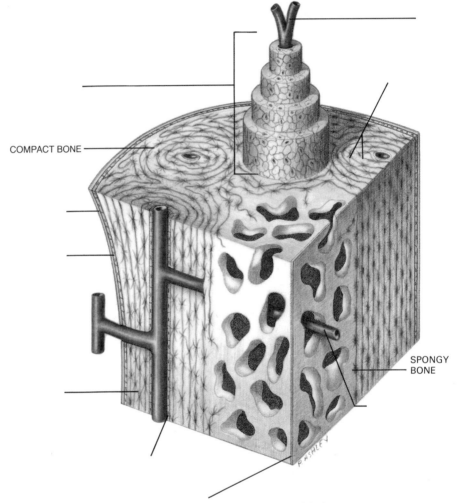

COMPACT BONE

SPONGY BONE

FIGURE 4-2 Fill in the correct labels.

V. THE SKULL

Anterior
Coronal
Cranial
Eight
Facial
14

Frontal
Lambdoidal suture
Middle
Paranasal
Parietal

Sagittal suture
Sinuses
Sinusitis
Soft spots
Sutures

Copyright © 1992 W. B. SAUNDERS COMPANY All rights reserved

1. The skull, the bony framework of the head, is divided into the _____

   and the _____ bones.

2. The cranium consists of _____ cranial bones that enclose the brain.

3. _____ bones make up the facial portion of the skull.

4. Contained within the head are six very small bones in the _____ ears.

5. Most of the bones of the skull are joined by the immovable joints called

   _____ .

6. The _____ is the joint between the two parietal bones.

7. The coronal suture joins the parietal bones to the _____ bone.

8. The _____ is the joint between the parietal bones and the occipital
   bone.

9. In babies, six joints called fontanelles occur at the angles of the _____
   bone.

10. The largest fontanelle is the _____ fontanelle, and it is found at the

    junction of the sagittal, frontal, and _____ sutures.

11. The fontanelles, popularly called _____ , permit the baby's head to be
    compressed slightly as it passes through the bony pelvis during birth.

12. _____ are air spaces lined with mucous membrane, and they are
    found in some of the cranial bones.

13. Four pairs of sinuses, the _____ sinuses, are continuous with the nose
    and throat.

14. Sometimes the mucous membranes of the sinuses become swollen and inflamed and

    produce the condition called _____ .

VI. THE VERTEBRAL COLUMN

Centrum	Five	Spine
Cervical	Fused	Spinous
Coccygeal	Intervertebral discs	Thoracic
Coccyx	Sacrum	24

1. The vertebral column, or _____ , supports the body and bears its weight.

2. The vertebral column consists of _____ vertebrae.

3. The two fused bones of the vertebral column are the _____ and the

   _____ .

4. The regions of the vertebral column are the _____ , which consists of

   seven vertebrae; the _____ , which consists of 12 vertebrae; the lumbar,

   which consists of _____ vertebrae; the sacral, which consists of five

   _____ vertebrae; and the _____ , which consists of
   fused vertebrae.

Copyright © 1992 W. B. SAUNDERS COMPANY All rights reserved

5. The vertebrae articulate with each other by means of synovial joints and by means of

_____ composed of cartilage.

6. The _____ is the bony central part of the vertebra that bears most of the body weight.

7. The _____ process is the posterior projection from the lamina, to which the back muscles are attached.

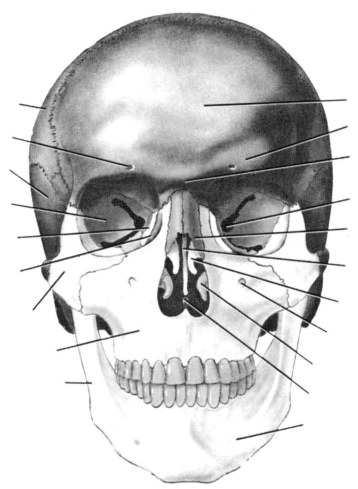

FIGURE 4–3 Fill in the correct labels.

Copyright © 1992 W. B. SAUNDERS COMPANY All rights reserved

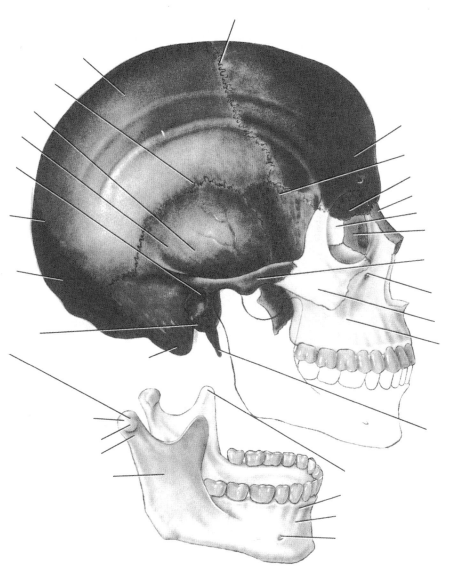

FIGURE 4–4 Fill in the correct labels.

Copyright © 1992 W. B. SAUNDERS COMPANY All rights reserved

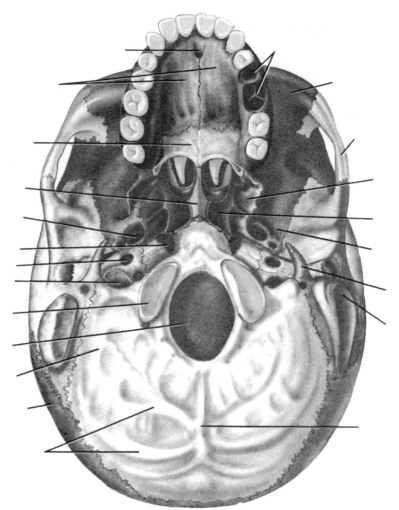

FIGURE 4–5 Fill in the correct labels.

VII. THE THORACIC CAGE

Pectoral   Sternum   12
Rib   Thoracic

1. The thoracic cage, or _____ cage, protects the internal organs of the chest, including the heart and lungs.

2. The thoracic cage provides support for the bones of the _____ girdle and upper limbs.

3. The thoracic cage is a bony cage formed by the _____, the _____ vertebrae, and _____ pairs of ribs.

Copyright © 1992 W. B. SAUNDERS COMPANY All rights reserved

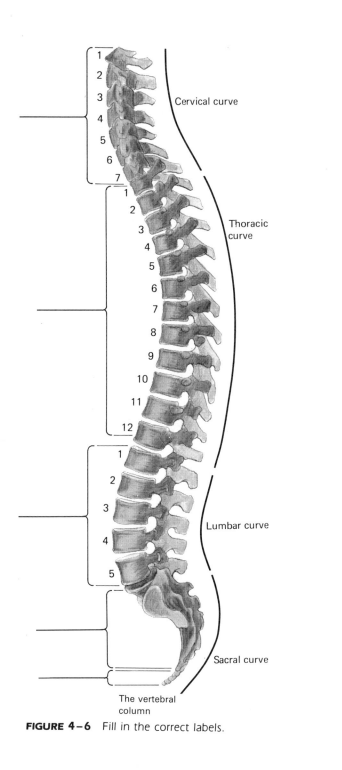

Cervical curve

Thoracic curve

Lumbar curve

Sacral curve

The vertebral column

FIGURE 4–6 *Fill in the correct labels.*

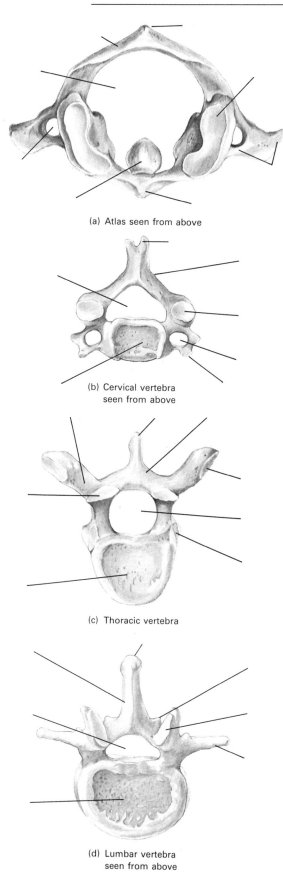

(a) Atlas seen from above

(b) Cervical vertebra seen from above

(c) Thoracic vertebra

(d) Lumbar vertebra seen from above

FIGURE 4–7 *Fill in the correct labels.*

Copyright © 1992 W. B. SAUNDERS COMPANY All rights reserved

VIII. THE PECTORAL AND PELVIC GIRDLES

Clavicle	Pectoral girdle	Sacrum
Coccyx	Pelvic girdle	Sternum
Coxal	Pelvis	Trunk
Inlet	Pubic	

1. The _____, or shoulder girdle, attaches the upper limbs to the axial skeleton.

2. Each pectoral girdle consists of a scapula and a(n) _____.

3. The pectoral girdles articulate with the _____ but not with the vertebral column.

4. The _____ is a broad basin of bone that encloses the pelvic cavity.

5. The pelvic girdle supports the lower limbs and is the site of attachment of the major muscles of the _____ and lower limbs.

6. Two _____ bones together with the _____ and the _____ form the pelvic girdle.

7. The female _____ is adapted for holding a developing baby and permitting its passage to the outside world at birth.

8. The pelvic _____ in the female is larger and more circular than that in the male.

9. In the female, there is a greater angle between the _____ bones.

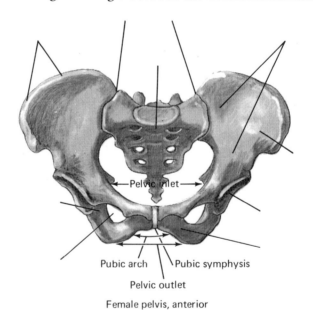

Female pelvis, anterior

FIGURE 4–8   *Fill in the correct labels.*

IX. THE UPPER AND LOWER LIMBS

Carpal	Metatarsals	Tarsal
Femur	Patella	30
Fibula	Phalanges	Tibia
Humerus	Radius	Ulna
Metacarpals		

Copyright © 1992 W. B. SAUNDERS COMPANY All rights reserved

1. The upper and lower limbs consist of _____ bones each.

2. The _____ is the bone in the upper arm.

3. The _____ and the _____ are the bones in the forearm.

4. The wrist bones are called the _____ bones.

5. The bones in the palm of the hand are called the _____.

6. The _____ are the bones in the fingers and the toes.

7. The upper leg or thigh bone is called the _____.

8. The _____ is the kneecap.

9. The _____ and the _____ are the bones in lower leg, or shin.

10. The bones in the back of the foot and heel are the _____ bones.

11. The _____ are the bones in the main part of the foot.

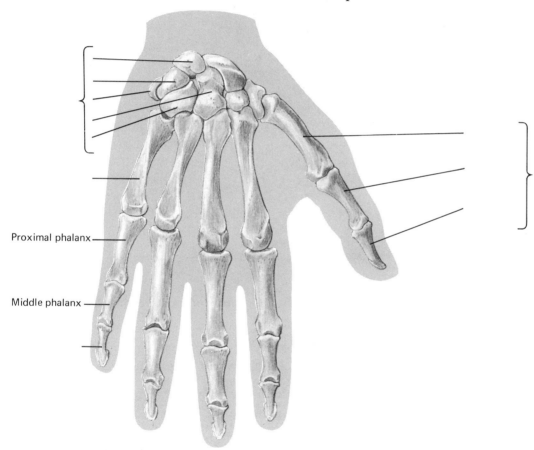

Proximal phalanx

Middle phalanx

Dorsal

FIGURE 4–9 Fill in the correct labels.

X. THE JOINTS

Amphiarthroses	Cartilage	Movement
Articulation	Connective	Synovial
Bursae	Diarthroses	Synovial fluid
Bursitis	Joints	Synthroses

1. A joint, or _____, is the point of contact between two bones.

Copyright © 1992 W. B. SAUNDERS COMPANY All rights reserved

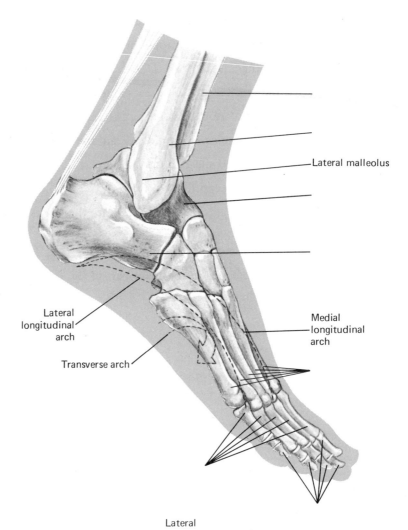

Lateral malleolus

Lateral
longitudinal
arch

Medial
longitudinal
arch

Transverse arch

Lateral

FIGURE 4-10 *Fill in the correct labels.*

2. _____ hold bones together, and many of them permit flexibility and movement.

3. Joints are classified according to the degree of _____ they permit.

4. Joints classified as _____ do not permit movement. These joints connect bones by means of fibrous _____ tissue.

5. Joints classified as _____ permit slight movement. With this type of joint, bones are joined by _____.

6. Diarthroses, or _____ joints, are referred to as freely movable joints.

7. Most of the body's joints are classified as _____ joints.

8. The joint capsule is lined with a membrane that secretes a lubricating

_____.

9. Fluid-filled sacs called _____ are located between bone and tendons and between bone and some other tissues.

10. Inflammation of a bursa is a painful condition called _____.

Copyright © 1992 W. B. SAUNDERS COMPANY All rights reserved

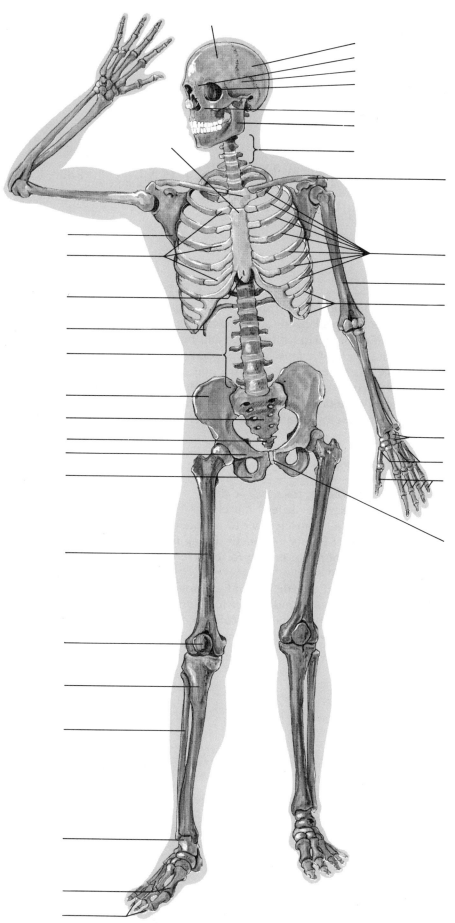

FIGURE 4-11 Fill in the correct labels.

Copyright © 1992 W. B. SAUNDERS COMPANY All rights reserved

Five

THE MUSCULAR SYSTEM

OUTLINE

I. Each skeletal muscle is an organ
II. Muscle fibers are specialized for contraction
III. Muscle contraction occurs when actin and myosin filaments slide past each other
IV. ATP provides energy for muscle contraction
V. Muscle tone is a state of partial contraction
VI. Two types of contraction are isotonic and isometric contraction
VII. Muscles work antagonistically to one another
VIII. We can study muscles in functional groups

LEARNING OBJECTIVES

After you have studied this chapter, you should be able to:
1. Describe the structure of a skeletal muscle.
2. Relate the structure of a muscle fiber to its function.
3. Trace the sequence of events that occur during muscle contraction.
4. Identify ATP as the source of energy for muscle contraction and identify the source of energy for making ATP.
5. Define muscle tone and explain why muscle tone is important.
6. Distinguish between istonic and isometric contraction.
7. Explain how muscles work antagonistically to one another.
8. Locate and give the actions of the principal muscles as indicated in Table 5–1 in the accompanying textbook.

Copyright © 1992 W. B. SAUNDERS COMPANY All rights reserved

NAME _____

STUDY QUESTIONS

Within each category, fill in the blanks with the correct response.

I. INTRODUCTION

Cardiac   Skeletal   Voluntary
Muscles   Smooth

1. All body movements depend on the action of _____.

2. The three types of muscles are _____, _____, and

   _____.

3. Skeletal muscles are the _____ muscles attached to bones.

II. EACH SKELETAL MUSCLE IS AN ORGAN

Endomysium   Fascicles   Perimysium
Epimysium   Fibers   Tendons

1. A skeletal muscle is an organ composed of hundreds of muscle cells, or

   _____.

2. Each muscle is surrounded by a covering of connective tissue called the

   _____.

3. The muscle fibers are arranged in bundles called _____.

4. Each fascicle is wrapped by connective tissue, the _____.

5. Individual muscle fibers are surrounded by a connective tissue covering, the

   _____.

6. Extensions of epimysium form tough cords of connective tissue, the

   _____, that anchor muscles to bones.

III. MUSCLE FIBERS ARE SPECIALIZED FOR CONTRACTION

Actin   Filaments   Nuclei
Contractile   Myosin   Transverse

1. Each muscle fiber is a spindle-shaped cell with many _____.

2. The plasma membrane has many inward extensions that form a set of

   _____ tubules.

3. Each muscle fiber is almost entirely composed of tiny protein threads, or

   _____.

Copyright © 1992 W. B. SAUNDERS COMPANY All rights reserved

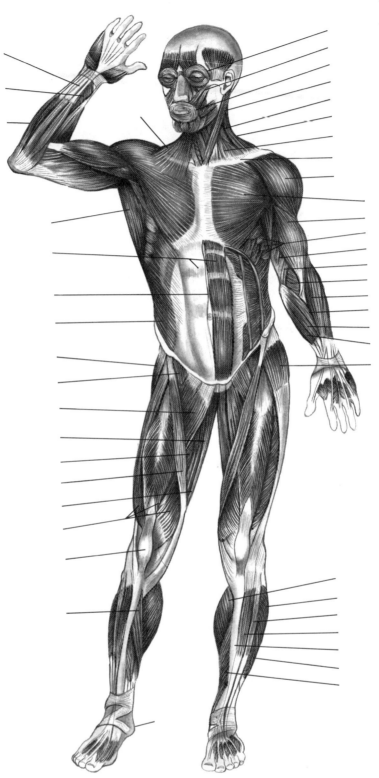

FIGURE 5-1   Fill in the correct labels.

Copyright © 1992 W. B. SAUNDERS COMPANY All rights reserved

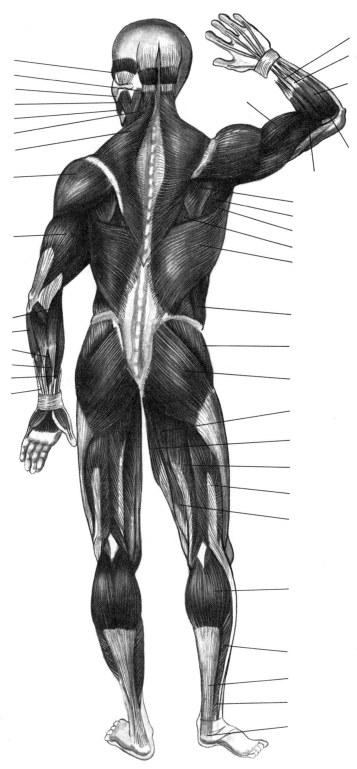

FIGURE 5–2 Fill in the correct labels.

Copyright © 1992 W. B. SAUNDERS COMPANY All rights reserved

4. The thick filaments consisting mainly of the protein myosin are called

   _____ filaments.

5. The thin filaments consisting of the protein actin are called _____
   filaments.

6. Myosin and actin are _____ proteins, which means that they are
   capable of shortening.

IV. MUSCLE CONTRACTION OCCURS WHEN ACTIN AND MYOSIN FILAMENTS SLIDE PAST EACH OTHER

Acetylcholine	Cholinesterase	Muscle
Actin	Fibers	Myosin
Action potential	Impulses	Neuromuscular
Bones	Motor	Receptors
Calcium		

1. Body movement occurs when muscles pull on _____.

2. A muscle contracts when its _____ contract.

3. Muscle fibers contract when the _____ and _____
   filaments actively pull themselves past and between one another.

4. A(n) _____ nerve is a nerve that controls muscle contraction.

5. Nerve fibers (neurons) from a motor nerve transmit _____ to muscle
   fibers.

6. The junction of a nerve and muscle fiber is called a(n) _____ junction.

7. A motor neuron releases a compound called _____.

8. Acetylcholine diffuses across the neuromuscular junction and combines with

   _____ on the surface of the muscle cell.

9. The electrical impulse generated in a muscle cell is called a(n) _____.

10. Excess acetylcholine is broken down by an enzyme called _____.

11. The action potential spreads through the T tubules and stimulates the release of

    _____.

12. The calcium stimulates the myosin and actin filaments to slide past each other in

    such a way that the _____ shortens.

V. ATP PROVIDES ENERGY FOR MUSCLE CONTRACTION

ATP	Glucose	Muscle fatigue
Creatine phosphate	Lactic acid	Oxygen debt
Fuel		

1. The immediate source of energy for muscle contraction comes from the energy

   storage molecule, _____.

Copyright © 1992 W. B. SAUNDERS COMPANY All rights reserved

2. In addition to ATP, muscle cells also have an energy storage compound called

   _____.

3. The energy for making creatine phosphate and ATP comes from _____
   molecules.

4. _____, a simple sugar, is stored in muscle cells in the form of a large
   molecule called glycogen.

5. The depletion of ATP results in weaker contractions and _____.

6. A waste product called _____ is produced during anaerobic metabolism
   of glucose.

7. During muscle exertion, a(n) _____ develops.

VI. MUSCLE TONE IS A STATE OF PARTIAL CONTRACTION

Isometric   Motor nerve   Unconscious
Isotonic   Muscle tone   Upright

1. Even when they are not moving, muscles are in a state of partial contraction called

   _____.

2. Muscle tone is a(n) _____ process that helps keep muscles prepared
   for action.

3. Muscle tone is also responsible for helping the muscles of the abdominal wall hold

   the internal organs in place and for helping our muscles keep us _____.

4. When the _____ to a muscle is cut, the muscle becomes limp, or flaccid.

5. _____ contraction occurs when muscles shorten and thicken.

6. _____ contraction occurs when muscle length does not appreciably
   change but muscle tension increases.

VII. MUSCLES WORK ANTAGONISTICALLY TO ONE ANOTHER

Agonist   Fixators   Synergists
Antagonist   Insertion   Tendons
Articulates   Origin

1. Skeletal muscles produce movements by pulling on _____, which in
   turn pull on bones.

2. When a muscle contracts, it draws one bone toward or away from the bone with

   which it _____.

3. The attachment of a muscle to a less movable bone is called its _____.

4. The attachment of a muscle to a more movable bone is called its _____.

5. The muscle that contracts to produce a particular action is called the

   _____, or prime mover.

6. The muscle that produces the opposite movement is called the _____.

Copyright © 1992 W. B. SAUNDERS COMPANY All rights reserved

7. _____ stabilize joints so that undesirable movement does not occur.

8. _____ stabilize the origin of an agonist so that its force is fully directed onto the bone into which it inserts.

VIII. FUNCTIONS OF MUSCLES

Biceps   Gluteus maximus   Rectus abdominis
Diaphragm   Masseter   Trapezius
Gastrocnemius   Pectoralis

1. The _____ raises the jaw.

2. The _____ draws the shoulder upward.

3. The _____ compresses abdominal contents.

4. The _____ increases the volume of the chest cavity.

5. The _____ rotates the arm medially.

6. The _____ flexes the elbow.

7. The _____ extends and rotates the thigh.

8. The _____ flexes the foot.

Copyright © 1992 W. B. SAUNDERS COMPANY All rights reserved

Six

THE CENTRAL NERVOUS SYSTEM

OUTLINE

I. The nervous system consists of the central nervous system and the peripheral nervous system
II. Neurons and glial cells are the cells of the nervous system
III. Bundles of axons make up nerves
IV. Neural function includes reception, transmission, integration, and response
V. The human brain is the most complex mechanism known
  A. The medulla contains vital centers
  B. The pons is a bridge to other parts of the brain
  C. The midbrain contains centers for visual and auditory reflexes
  D. The diencephalon includes the thalamus and hypothalamus
  E. The cerebellum is responsible for coordination of movement
  F. The cerebrum is the largest part of the brain
    1. The cerebrum is divided into hemispheres
    2. The white matter of the cerebrum serves three functions
    3. The cerebrum has sensory, motor, and association functions
    4. The lobes of the cerebrum specialize in specific functions
VI. The spinal cord transmits information to and from the brain
VII. The central nervous system is well protected
  A. The meninges are connective tissue coverings
  B. Cerebrospinal fluid cushions the central nervous system
VIII. A reflex action is a simple neural response

LEARNING OBJECTIVES

After you have studied this chapter, you should be able to:
1. List the divisions of the nervous system.
2. Draw a neuron, label its parts, and give the function of each part.
3. Distinguish between nerve and tract and between ganglion and nucleus.
4. Briefly describe the four basic processes on which all neural responses depend — reception, transmission, integration, and response.
5. Label a diagram of the structures of the brain described in this chapter.
6. Describe the structure and functions of the main parts of the brain — medulla, pons, midbrain, diencephalon (thalamus and hypothalamus), cerebellum, and cerebrum.
7. Name the principal areas and functions associated with the lobes of the cerebrum.
8. List two functions of the spinal cord and describe its structure.
9. Describe the structures that protect the brain and spinal cord.
10. Diagram a withdrawal reflex, identifying the essential structures and indicating the direction of impulse transmission.

Copyright © 1992 W. B. SAUNDERS COMPANY All rights reserved

NAME _____

STUDY QUESTIONS

Within each category, fill in the blanks with the correct response.

I. THE NERVOUS SYSTEM CONSISTS OF THE CENTRAL NERVOUS SYSTEM AND THE PERIPHERAL NERVOUS SYSTEM

Afferent	Cranial	Sense
Autonomic	Efferent	Somatic
Brain	Homeostasis	Spinal
Central	Peripheral	

1. The nervous system works continuously with the endocrine system to maintain

   _____ .

2. Two principal divisions of the nervous system are the _____ nervous

   system and the _____ nervous system.

3. The central nervous system consists of the _____ and spinal cord.

4. The peripheral nervous system is made up of _____ receptors and the
   nerves that are the communication lines to and from the central nervous system.

5. Twelve pairs of _____ nerves link the brain and 31 pairs of

   _____ nerves link the spinal cord with sense organs, muscles, and
   other parts of the body.

6. The peripheral nervous system may be subdivided into _____ and

   _____ divisions.

7. _____ nerves transmit messages from receptors to the central nervous
   system.

8. _____ nerves transmit information back from the central nervous
   system to the structures that must respond.

II. NEURONS AND GLIAL CELLS ARE THE CELLS OF THE NERVOUS SYSTEM

Axon	Glial	Neurilemma
Cellular	Impulses	Neurons
Dendrites	Myelin	Neurotransmitters
Fibers		

1. _____ cells protect and support the neurons.

2. _____ are cells that are highly specialized to receive and transmit
   messages in the form of neural impulses.

3. The neuron is distinguished from all other cells by its _____ .

Copyright © 1992 W. B. SAUNDERS COMPANY All rights reserved

4. _____ are multibranched fibers that project from the cell body and are highly specialized to receive neural impulses.

5. The _____ transmits neural messages from the cell body toward another neuron.

6. Synaptic knobs release _____, chemical substances that transmit impulses from one neuron to another.

7. Axons of many neurons of the peripheral nervous system are covered by two sheaths:

    an inner _____ sheath and an outer _____ sheath.

8. The _____ is important in the repair of injured neurons.

9. Myelin is an excellent electrical insulator that speeds the conduction of nerve

    _____.

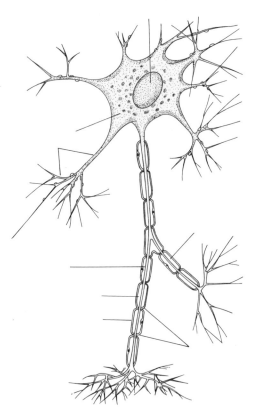

FIGURE 6–1 Fill in the correct labels.

III. BUNDLES OF AXONS MAKE UP NERVES

Axons   Myelin   Nuclei
Ganglion   Nerve   Tracts

1. A _____ is a large bundle of axons wrapped in connective tissue.

2. In comparing a nerve to a telephone cable, the _____ are like the

    individual wires, whereas the _____ sheaths are like the insulation.

Copyright © 1992 W. B. SAUNDERS COMPANY All rights reserved

3. The cell bodies attached to the axons of a nerve are often grouped together in a mass called a _____.

4. Within the central nervous system, bundles of axons are called _____ instead of nerves.

5. Within the central nervous system, masses of cell bodies are called _____ rather than ganglia.

IV. NEURAL FUNCTION INCLUDES RECEPTION, TRANSMISSION, INTEGRATION, AND RESPONSE

1. Number the following processes in order of occurrence from 1 to 5.

_____ Integration

_____ Reception

_____ Transmission (to the muscles)

_____ Transmission (to the central nervous system)

_____ Actual response

Dendrites   Neurotransmitters   Synaptic
Neurons   Synapse

2. Neural messages travel over sequences of _____.

3. Neurons are arranged so that the axon of one neuron forms junctions with the _____ of other neurons.

4. A junction between two neurons is called a _____.

5. At a synapse, neurons are separated by a tiny gap called the _____ cleft.

6. _____ conduct "messages" across the synaptic cleft.

V. THE HUMAN BRAIN IS THE MOST COMPLEX MECHANISM KNOWN

Brain   Medulla   Thalamus
Brainstem   Midbrain   Vasomotor
Cardiac   Oblongata   Ventricles
Cerebellum   Pons   Ventricles
Cerebrum   Stroke (cerebro-
Cortex   vascular accident)

1. _____ cells require a continuous supply of oxygen and glucose.

2. The most common cause of brain damage is a(n) _____.

3. The medulla, pons, and midbrain make up the _____.

4. The brain is a hollow organ; its fluid-filled spaces are called _____.

Copyright © 1992 W. B. SAUNDERS COMPANY All rights reserved

5. Formally known as the medulla _____, the medulla is the most posterior portion of the brainstem.

6. Among the vital centers of the medulla, the _____ centers control heart rate, and the _____ centers help regulate blood pressure by controlling blood vessel diameters.

7. The _____ forms a bulge on the anterior surface of the brainstem.

8. The pons consists mainly of nerve fibers passing between the _____ and other parts of the brain.

9. The _____ is the shortest portion of the brainstem. It extends from the pons to the diencephalon.

10. The _____ consists of two oval masses, one located on each side of the third ventricle. It functions as a major relay center.

11. The second largest part of the brain is the _____. It consists of two lateral masses called hemispheres and a connecting portion.

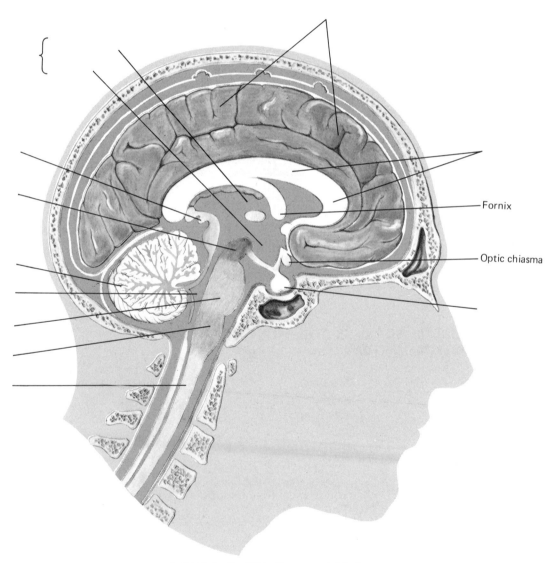

Fornix

Optic chiasma

FIGURE 6–2 Fill in the correct labels.

Copyright © 1992 W. B. SAUNDERS COMPANY All rights reserved

12. The _____ is the largest and most prominent part of the human brain.

13. The thin outer layer of the cerebrum consists of gray matter and is called the cerebral

_____ .

14. The two cavities within the cerebrum are the lateral _____ .

VI. THE SPINAL CORD TRANSMITS INFORMATION TO AND FROM THE BRAIN

Ascending	Fissures	Vertebral
Descending	Spinal	

1. The _____ cord has two main functions: (1) it controls many reflex activities of the body, and (2) it transmits information between the nerves and the peripheral nervous system of the brain.

2. The spinal cord occupies the _____ canal of the vertebral column.

3. Several grooves called _____ divide the spinal cord into regions.

4. _____ tracts transmit sensory information up the spinal cord to the brain.

5. _____ tracts transmit impulses from the brain down the spinal cord to the efferent nerves.

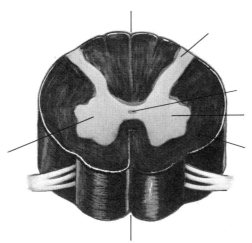

FIGURE 6-3 Fill in the correct labels.

VII. THE CENTRAL NERVOUS SYSTEM IS WELL PROTECTED

Arachnoid	Meninges	Sinuses
Cerebrospinal	Meningitis	Spinal
Dura mater	Pia mater	
Encephalitis	Reflex	

1. The three connective tissue layers covering the brain and spinal cord are the

_____ .

Copyright © 1992 W. B. SAUNDERS COMPANY All rights reserved

2. The outermost layer of the meninges is the _____, a tough, double-layered membrane.

3. The second layer of the meninges is the _____, a thin, delicate membrane.

4. The innermost layer of the meninges is the _____, a very thin membrane that adheres closely to the brain and spinal cord.

5. Inside the skull, the two layers of the dura mater are separated in some regions by large blood vessels called _____. These vessels receive blood leaving the brain and deliver it to the jugular veins in the neck.

6. _____ is an inflammation of the meninges.

7. Viral meningitis that spreads and causes inflammation of the brain is called _____.

8. The shock-absorbing fluid that fills the ventricles (the cavities within the brain) and the spaces below the arachnoid layer in the brain and spinal cord is called _____ fluid.

9. A(n) _____ tap can be used to withdraw small amounts of cerebrospinal fluid without damaging the spinal cord.

10. A simple example of a neural response is a(n) _____ action.

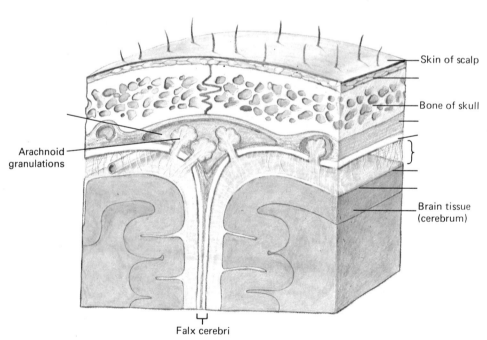

Skin of scalp

Bone of skull

Arachnoid granulations

Brain tissue (cerebrum)

Falx cerebri

FIGURE 6–4 Fill in the correct labels.

Copyright © 1992 W. B. SAUNDERS COMPANY All rights reserved

Seven

THE PERIPHERAL NERVOUS SYSTEM

OUTLINE

I. The somatic system responds to changes in the outside world
   A. Cranial nerves link the brain with sense receptors and muscles
   B. Spinal nerves link the spinal cord with various structures
      1. Each spinal nerve divides into branches
      2. The ventral branches form plexuses
II. The autonomic system maintains internal balance
   A. The sympathetic system mobilizes energy
   B. The parasympathetic system conserves and restores energy
   C. Sympathetic and parasympathetic nerves have opposite effects on many organs

LEARNING OBJECTIVES

After you have studied this chapter, you should be able to:
1. Contrast the somatic with the autonomic divisions of the peripheral nervous system.
2. List the cranial nerves and give the functions of each.
3. Describe the structure of a typical spinal nerve.
4. Name and describe the major plexuses.
5. Describe the structures of a reflex pathway in the autonomic system.
6. Compare and contrast the sympathetic system with the parasympathetic system.
7. Compare the effect of sympathetic stimulation with that of parasympathetic stimulation on specific organs such as the heart and the digestive tract.

Copyright © 1992 W. B. SAUNDERS COMPANY All rights reserved

NAME _____

STUDY QUESTIONS

Within each category, fill in the blanks with the correct response.

I. THE SOMATIC SYSTEM RESPONDS TO CHANGES IN THE OUTSIDE WORLD

Autonomic	Dorsal	Spinal
Cervical	Sense	Ventral
Cranial	Somatic	

1. The peripheral nervous system is made up of _____ receptors, the nerves that link the sense organs with the central nervous system, and the nerves that link the central nervous system with the muscles and glands.

2. The portion of the peripheral nervous system that keeps the body in adjustment with the outside world is the _____ system.

3. The nerves and receptors that maintain internal balance make up the _____ system.

4. Like those of the autonomic system, the afferent and efferent neurons of the somatic system are part of the _____ and _____ nerves.

5. Each spinal nerve has two points of attachment with the cord: the _____ root and the _____ root.

6. The _____ plexus receives sensory information from the back of the head, neck, shoulders, and upper chest.

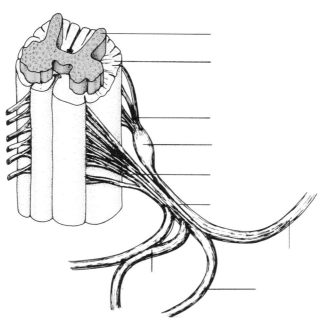

FIGURE 7–1 Fill in the correct labels.

Copyright © 1992 W. B. SAUNDERS COMPANY All rights reserved

II. THE AUTONOMIC SYSTEM MAINTAINS INTERNAL BALANCE

Autonomic   Efferent   Sympathetic
Brain   Parasympathetic   Vagus
Conserves   Pelvic

1. The _____ system works to maintain a steady state within the body.

2. The efferent portion of the autonomic system is subdivided into _____

   and _____ systems.

3. In the autonomic system, two _____ neurons are found between the central nervous system and the muscle it innervates.

4. The parasympathetic system _____ energy.

5. Neurons of the parasympathetic system emerge from the _____ and from the sacral region of the spinal cord.

6. Most of the parasympathetic fibers are in the _____ nerves.

7. The parasympathetic nerves that emerge from the sacral region form the

   _____ nerves.

Copyright © 1992 W. B. SAUNDERS COMPANY All rights reserved

Eight

THE SENSE ORGANS

OUTLINE

I. The eye contains visual receptors
   A. The retina contains the rods and cones
   B. The eye can be compared to a camera
   C. Vision involves several processes
II. The ear functions in hearing and equilibrium
   A. The outer ear conducts sound waves to the middle ear
   B. The middle ear amplifies sound waves
   C. The inner ear contains receptors
      1. The cochlea contains the receptors for hearing
      2. The vestibule and semicircular canals help maintain equilibrium
III. Smell is sensed by receptors in the nasal cavity
IV. Taste is sensed by the taste buds
V. The general senses are widespread throughout the body
   A. Tactile receptors are located in the skin
   B. Temperature receptors are nerve endings
   C. Pain sensation is a protective mechanism
   D. Proprioceptors inform us of our position

LEARNING OBJECTIVES

After you have studied this chapter, you should be able to:
1. Describe the structures of the eye and give their functions.
2. Describe the structures and functions of the three major parts of the ear.
3. Trace the transmission of sound through the ear.
4. Compare the receptors of taste with those of smell.
5. Describe the tactile and temperature receptors.
6. Explain how pain sensation is protective and describe referred pain.
7. Locate proprioceptors in the body and describe their function.

Copyright © 1992 W. B. SAUNDERS COMPANY All rights reserved

NAME _____

STUDY QUESTIONS

Within each category, fill in the blanks with the correct response.

I. INTRODUCTION

Hearing	Smell	Touch
Receptors	Stimulus	
Sight	Taste	

1. Any detectable change in the environment is called a _____ .

2. Traditionally, the five senses have been referred to as _____ ,

   _____ , _____ , _____ , and

   _____ .

3. How we respond to changes in our environment depends on _____
   that sense changes in the outside world.

II. THE EYE CONTAINS VISUAL RECEPTORS

Blinking	Fovea	Orbit
Cones	Iris	Pupil
Conjunctiva	Lacrimal	Reflex
Cornea	Lashes	Rods
Ducts	Lids	Sclera
Extrinsic	Optic	Sensory
Fat	Optic disc	

1. The eye and its muscles are set in the _____ formed by the skeletal
   bones of the face.

2. The eye and its muscles are cushioned by layers of _____ .

3. The eye _____ and _____ help protect the eye
   anteriorly from foreign objects.

4. The eyelids close by _____ action if danger is perceived.

5. Frequent _____ of the eye lubricates the eye and clears debris.

6. Tears are excreted at all times from the _____ glands.

7. Tears pass through the lacrimal _____ to keep the eye moist and free
   of dust and minute objects.

8. The six _____ muscles of the eye originate from outside the eye and
   function in support and movement.

9. The _____ is called the "white" of the eye.

10. The _____ is frequently called the "window" of the eye.

11. The sclera is covered by the _____ , a moist mucous membrane that
    extends as a continuous lining of the inner layer of the eyelids.

Copyright © 1992 W. B. SAUNDERS COMPANY All rights reserved

12. The _____ is the colored part of the eye.

13. The black spot in the middle of the eye is called the _____ .

14. The retina contains _____ receptors called rods and cones.

15. The _____ are mainly responsible for color vision and vision during daytime.

16. The _____ are mainly responsible for vision in dim light or darkness.

17. Cones are most concentrated in the _____, a small depression in the center of the posterior region of the retina.

18. The image of an object is carried from the retina to the brain through the

_____ nerve.

19. In the area where the optic nerve forms, there are no rods or cones. This area is

called the _____, or blind spot.

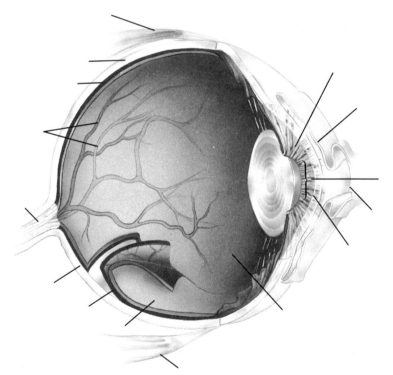

FIGURE 8–1   Fill in the correct labels.

III. THE EAR FUNCTIONS IN HEARING AND EQUILIBRIUM

Cerumen	Membranous	Temporal
Cochlea	Middle	Tympanic
Corti	Perilymph	Vestibule
Endolymph	Pinna	
Labyrinth	Semicircular	

1. The _____ is the part of the ear that projects from the side of the head and surrounds the ear canal.

Copyright © 1992 W. B. SAUNDERS COMPANY All rights reserved

2. The ear canal leads to the _____ ear.

3. _____, commonly called earwax, helps protect the lining of the ear canal from infection.

4. The _____ membrane, commonly called the eardrum, separates the middle and external ears.

5. The inner ear lies inside the _____ bone.

6. The inner ear is a bony _____ composed of three compartments.

7. The bony labyrinth contains a fluid called _____.

8. The _____ labyrinth is a group of ducts and sacs that lie within the bony labyrinth.

9. The membranous labyrinth contains a fluid called _____.

10. The _____ is a snail-shaped portion of the middle ear.

11. The cochlea contains the organs of _____, the sound receptors.

12. The _____ and _____ canals make up the "organ of equilibrium."

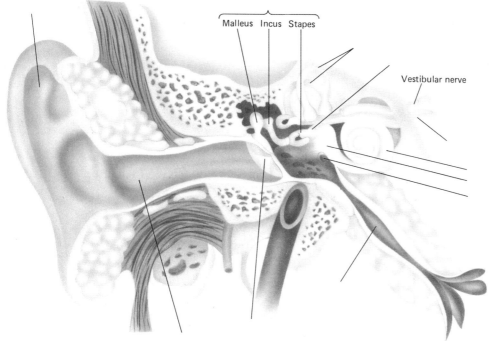

FIGURE 8-2 Fill in the correct labels.

IV. SMELL IS SENSED BY RECEPTORS IN THE NASAL CAVITY

Nerve   Olfactory   Smell

1. Smell is sensed by receptor cells in the _____ epithelium located at the upper part of the nasal cavity.

2. Smells are transmitted to the olfactory center in the brain through the olfactory

_____.

Copyright © 1992 W. B. SAUNDERS COMPANY All rights reserved

3. Olfactory cells are not replaced so when they are damaged, our sense of

_____ is impaired.

V. TASTE IS SENSED BY THE TASTE BUDS

Appetite	Sides	Tongue
Back	Solution	
Decreases	Tip	

1. Taste buds are found on the _____ and various parts of the mouth.

2. As an individual gets older, the number of taste buds gradually _____.

3. Taste buds respond only when the material to be tasted is in _____.

4. Sweet and salty tastes are sensed at the _____ of the tongue.

5. Sour taste is detected mainly at the _____ of the tongue.

6. Bitter taste is detected mainly at the _____ of the tongue.

7. Both taste and smell are important in stimulating _____ and digestive juices.

VI. THE GENERAL SENSES ARE WIDESPREAD THROUGHOUT THE BODY

Acupuncture	Pain	Referred
Anus	Pain	Tactile
Lips	Proprioceptors	Tissue
Mouth	Radiate	

1. _____ receptors respond to pressure, touch, and vibration.

2. Temperature receptors are widely distributed and are particularly concentrated in the

_____, _____, and _____.

3. The sensation of _____ is a protective mechanism that makes us aware

of _____ injury.

4. _____ from internal organs is often difficult to locate.

5. _____ pain occurs when the pain is felt in an area different from that in which it originates.

6. When internal pain is felt both at the site of distress and as referred pain, it may

appear to spread, or _____.

7. _____ is one of the oldest techniques used for relieving pain.

8. _____ help us maintain the position of the body and its parts.

Copyright © 1992 W. B. SAUNDERS COMPANY All rights reserved

16. Sweet and salty tastes are sensed mainly at _____.
    a. The tip of the tongue
    b. The sides of the tongue
    c. The back of the tongue
    d. Under the tongue

17. Sour taste is detected mainly at _____.
    a. The tip of the tongue
    b. The sides of the tongue
    c. The back of the tongue
    d. Under the tongue

18. Bitter taste is detected mainly at _____.
    a. The tip of the tongue
    b. The sides of the tongue
    c. The back of the tongue
    d. Under the tongue

Copyright © 1992 W. B. SAUNDERS COMPANY All rights reserved

Nine

ENDOCRINE CONTROL

OUTLINE

I. Hormones act on target cells
  A. Many hormones act through second messengers
  B. Steroid hormones activate genes
  C. Prostaglandins are local hormones
II. The endocrine glands are regulated by feedback control
III. The hypothalamus and the pituitary gland work closely together
  A. The posterior lobe releases two important hormones
  B. The anterior lobe secretes seven different hormones
    1. Tropic hormones stimulate other endocrine glands
    2. Prolactin stimulates secretion of milk
    3. Growth hormone stimulates protein synthesis
IV. The thyroid gland is located in the neck
V. The parathyroid glands are located on the thyroid
VI. The islets of Langerhans are the endocrine portion of the pancreas
  A. Insulin and glucagon regulate the concentration of glucose in the blood
  B. In diabetes mellitus, glucose accumulates in the blood
VII. The adrenal glands function in metabolism and stress
  A. The adrenal medulla synthesizes epinephrine and norepinephrine
  B. The adrenal cortex secretes steroid hormones
VIII. Stress threatens homeostasis
IX. Several other tissues secrete hormones

LEARNING OBJECTIVES

After you have studied this chapter, you should be able to:
1. Define the term ''hormone'' and describe the functions of hormones.
2. Identify the principal endocrine glands and locate them in the body.
3. Describe how hormones work.
4. Describe how endocrine glands are regulated by feedback mechanisms.
5. Justify describing the hypothalamus as the link between the neural and endocrine systems. (Describe the mechanisms by which the hypothalamus exerts its control.)
6. Identify the hormones released by the anterior and posterior lobes of the pituitary, give their origins, and describe their actions.
7. Identify the hormones secreted by the thyroid gland and summarize their actions.
8. Describe how thyroid hormone secretion is regulated, and describe the effects of hyposecretion and hypersecretion.
9. Describe how the parathyroid and thyroid glands regulate calcium levels.

88

Copyright © 1992 W. B. SAUNDERS COMPANY All rights reserved

10. Contrast the actions of insulin with those of glucagon.
11. Describe the role of the adrenal medulla in the body's responses to stress. (Give the actions of epinephrine and norepinephrine.)
12. Identify the hormones secreted by the adrenal cortex, and give the actions of glucocorticoids and mineralocorticoids.

Copyright © 1992 W. B. SAUNDERS COMPANY All rights reserved

NAME _____

STUDY QUESTIONS

Within each category, fill in the blanks with the correct response.

I. INTRODUCTION

Endocrine   Hormones   Target
Endocrinology

1. The _____ system works with the nervous system to maintain the steady state of the body.

2. The endocrine system consists of tissues and glands that secrete chemical messengers called _____.

3. Hormones affect the activity of their _____ tissues, the tissues on which they act.

4. _____ is the study of endocrine function.

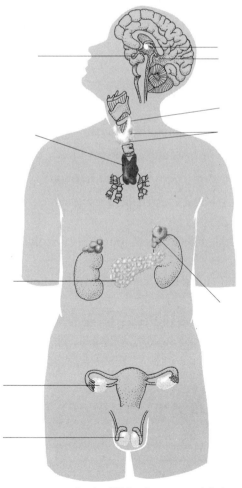

FIGURE 9–1 Fill in the correct labels.

Copyright © 1992 W. B. SAUNDERS COMPANY All rights reserved

II. HORMONES ACT ON TARGET CELLS

Local   Plasma   Proteins
Messenger   Prostaglandins   Receptor

1. Specialized _____ on or in the target cell act as receptors and bind hormones.

2. When the hormone combines with the _____, a series of reactions are activated.

3. Because the hormone turns on the system, it is referred to as the first

   _____ .

4. Hormones that are large molecules combine with receptors on

   the _____ membrane of the target cell.

5. _____ are a group of closely related lipids that are manufactured by many different tissues in the body.

6. Prostaglandins are sometimes referred to as _____ hormones because they act on nearby cells.

III. ENDOCRINE GLANDS ARE REGULATED BY FEEDBACK CONTROL

Feedback   Inhibited   Positive
Hypersecretion   Negative   Secretion
Hyposecretion   Parathyroid

1. Endocrine glands are regulated by _____ control.

2. _____ hormone helps regulate the calcium concentration in the blood.

3. When calcium concentration rises above normal limits, the parathyroid glands are

   _____ and slow their output of hormone.

4. In _____ feedback mechanisms, the effects are opposite those of the stimuli.

5. In _____ feedback, the output increases the hormone secretion rather than halting it.

6. When an endocrine gland is not regulated effectively, the rate of _____ becomes abnormal.

7. In _____, a gland decreases its hormone output.

8. In _____, a gland increases its hormone output to abnormal levels and overstimulates target cells.

IV. THE HYPOTHALAMUS AND PITUITARY GLAND WORK TOGETHER CLOSELY

ADH   Hypothalamus   Posterior
Anterior   Oxytocin   Protein
Hormones   Pituitary

Copyright © 1992 W. B. SAUNDERS COMPANY All rights reserved

1. Endocrine activity is controlled by the _____, which links the nervous and endocrine systems.

2. The hypothalamus secretes several releasing and inhibiting _____.

3. The _____ gland is sometimes called the master gland of the body.

4. The _____ lobe of the pituitary gland secretes two hormones, oxytocin and ADH.

5. _____ stimulates contraction of smooth muscle in the wall of the uterus.

6. _____ regulates fluid balance in the body and indirectly helps control blood pressure.

7. The _____ lobe of the pituitary gland secretes six important hormones, including growth hormone.

8. Growth hormone stimulates body growth, mainly by stimulating _____ synthesis.

V. THE THYROID GLAND IS LOCATED IN THE NECK

Goiter	Isthmus	Thyroxine
Hypothyroidism	Metabolic	TSH
Iodine	Thyroid	

1. Shaped somewhat like a shield, the _____ gland is located in the neck.

2. The two lobes of the thyroid gland are connected by a bridge of tissue called the

_____ .

3. The main thyroid hormone is _____ .

4. The thyroid hormones control _____ rate and promote growth.

5. The anterior pituitary secretes _____, which promotes synthesis and secretion of thyroid hormones.

6. Extreme _____ during childhood results in low metabolic rate and retarded mental and physical development.

7. Any abnormal enlargement of the thyroid gland is called a(n) _____ .

8. One cause of a goiter is a(n) _____ deficiency.

VI. PARATHYROID GLANDS ARE LOCATED ON THE THYROID

Calcitonin	Decreased	Parathyroid
Calcium	Increased	PTH

1. The _____ glands are embedded in the connective tissue that surrounds the thyroid gland.

2. The parathyroid glands secrete _____ .

3. The parathyroid glands are regulated by the concentration of _____ in the blood and tissue fluid.

Copyright © 1992 W. B. SAUNDERS COMPANY All rights reserved

4. When calcium concentrations become very high, _____ is released from the thyroid gland.

5. When PTH is too low, the calcium level is _____.

6. When PTH is too high, the calcium level is _____.

VII. THE ISLETS OF LANGERHANS ARE THE ENDOCRINE PORTION OF THE PANCREAS

Alpha	Glycogen	Mellitus
Beta	Hyperglycemic	Noninsulin
Decrease	Insulin	Pancreas
Glucagon	Langerhans	Sugar

1. The _____ lies in the abdomen posterior to the stomach and has both exocrine and endocrine functions.

2. More than 1 million small clusters of cells, called the islets of _____, are scattered throughout the pancreas.

3. Approximately 70% of the islet cells are _____ cells that produce the hormone insulin.

4. _____ cells secrete the hormone glucagon.

5. The main action of _____ is to stimulate the storage of glucose.

6. Cells store glucose by converting it into _____.

7. _____ raises the blood sugar level.

8. Secretion of insulin and glucagon is directly controlled by the blood _____ level.

9. The main disorder associated with pancreatic hormones is diabetes _____.

10. In insulin-dependent diabetes, there is a marked _____ in the number of beta cells in the pancreas.

11. More than 90% of all cases of diabetes are _____ dependent, or type II.

12. When blood glucose levels rise above the normal range, the individual is said to be _____.

VIII. THE ADRENAL GLANDS FUNCTION IN METABOLISM AND STRESS

ACTH	Cortisol	Mineralocorticoids
Adrenal	Epinephrine	Norepinephrine
Adrenalin	Glucocorticoids	Sex
Aldosterone	Glucose	Steroids
Cortex	Medulla	Sympathetic

1. The paired _____ glands are small yellow masses of tissue located above the kidneys.

Copyright © 1992 W. B. SAUNDERS COMPANY All rights reserved

2. Each gland consists of a central portion, the adrenal _____, and a larger outer region, the adrenal _____.

3. The adrenal medulla develops from nervous tissue and is sometimes considered part of the _____ nervous system.

4. The adrenal medulla secretes two hormones, _____ and _____.

5. Epinephrine is commonly called _____.

6. The adrenal cortex secretes three different types of hormones, which are all _____.

7. _____ help the body cope with stress.

8. The main glucocorticoid is _____.

9. The principal action of glucocortoids is to promote production of _____ from other nutrients.

10. _____ are the second type of hormone secreted by the adrenal cortex. They regulate the balance of water and salt.

11. _____ is the principal mineralocorticoid.

12. _____ hormones are also secreted by the adrenal cortex, but the amounts are so small that they have little effect on the body.

13. Secretion of hormones by the adrenal cortex is regulated by _____, which is secreted from the pituitary.

IX. STRESS THREATENS HOMEOSTASIS

Alarm	Homeostasis	Stress
Exhaustion	Resistance	Stressors

1. Good health and survival depend on maintaining _____.

2. _____ are stimuli that disrupt the steady state of the body.

3. General adaptation syndrome (GAS) is a term used to describe the body's response to _____.

4. In the first stage of GAS, the _____ reaction, the adrenal medulla prepares the body for fight or flight.

5. The _____ reaction stage of GAS occurs if stress continues for a long period of time.

6. The final stage of GAS is _____, in which the body appears to wear out, and death may occur.

Copyright © 1992 W. B. SAUNDERS COMPANY All rights reserved

16. Alpha cells produce the hormone _____ .
    a. Glucagon
    b. ADH
    c. Insulin
    d. Oxytocin
    e. None of the preceding

17. The main disorder associated with pancreatic hormones is _____ .
    a. Diabetes insipidus
    b. Diabetes mellitus
    c. Pancreatic cancer
    d. All of the preceding
    e. None of the preceding

18. _____ is not a stage of General Adaptation Syndrome.
    a. Exhaustion
    b. Alarm reaction
    c. Resistance reaction
    d. Death
    e. All of the preceding

Copyright © 1992 W. B. SAUNDERS COMPANY All rights reserved

NAME _____

Crossword Puzzle for Chapters 6, 7, 8, and 9

ACROSS

1 Nerve that supplies the diaphragm
4 Second cranial nerve
8 Tropic hormone released by anterior pituitary gland
9 Part of nervous system consisting of brain and spinal cord
10 Focuses light on retina
12 Sensory fibers enter the spinal cord through the dorsal _____.
13 Opening through which light enters the eye
17 Part of the brain that regulates heart rate and blood pressure
20 Each of the six divisions of the cerebral hemisphere is called a _____.
21 Innermost of the three meninges
22 An action system of the brain
25 Part of the brain that controls body temperature
26 Releases growth hormone

DOWN

1 Controlled by the cerebellum
2 Sensitive to color
3 Hormone that regulates calcium level
5 Part of the brain that controls voluntary movement
6 Part of the brain that helps regulate respiration
7 Lobe where Broca's speech area is located
9 Gray matter of cerebrum
11 Visual and auditory reflex centers are located in the _____ brain.
12 Automatic sequence of stimulus response
14 Part of the nervous system that includes sense organs
15 Region of spinal cord below the thoracic level
16 Transmits impulses from the cell body toward the synapse
18 Hormone that stimulates reabsorption of water
19 Transmits impulses to cell body
23 Gland that stimulates metabolic rate
24 Outer layer of the meninges

Copyright © 1992 W. B. SAUNDERS COMPANY All rights reserved

Ten

THE CIRCULATORY SYSTEM: BLOOD

OUTLINE

    I. Blood consists of cells and platelets suspended in plasma
   II. Plasma is the fluid component of blood
  III. Red blood cells transport oxygen
  IV. White blood cells defend the body against disease
   V. Platelets function in blood clotting
  VI. Successful blood transfusions depend on blood groups
      A. The ABO system consists of antigens A and B
      B. The Rh system consists of several Rh antigens

LEARNING OBJECTIVES

After you have studied this chapter, you should be able to:
1. List the functions of the circulatory system.
2. Describe the composition of blood plasma and the functions of plasma proteins.
3. Describe the structure and function of red blood cells.
4. Describe the structure and function of white blood cells.
5. Describe the structure and function of platelets and summarize the chemical events of blood clotting.
6. Identify the antigen and antibody associated with each ABO blood type, and explain why blood types must be carefully matched in transfusion therapy.
7. Identify the cause and importance of Rh incompatibility.

Copyright © 1992 W. B. SAUNDERS COMPANY All rights reserved

NAME _____

STUDY QUESTIONS

Within each category, fill in the blanks with the correct response.

I. INTRODUCTION

Blood	Circulatory	Heart
Cardiovascular	Fluid	Lymphatic

1. The _____ system is the transportation system of the body.

2. Most of the substances transported by the circulatory system are carried by the

   _____.

3. The circulatory system consists of two subsystems: the _____ system

   and the _____ system.

4. In the cardiovascular system, the _____ pumps blood through a vast
   system of blood vessels.

5. The lymphatic system helps preserve _____ balance and protects the
   body against disease.

II. BLOOD CONSISTS OF CELLS AND PLATELETS SUSPENDED IN PLASMA

Alkaline	Platelets	6
5.6	Red	White
Plasma		

1. Blood consists of _____ blood cells and _____ blood
   cells.

2. In addition to red blood cells and white blood cells, blood also contains cell fragments

   called _____.

3. Blood cells and platelets are suspended in a pale-yellow fluid called

   _____.

4. In an average size adult, blood volume is normally approximately

   _____ l, or _____ qt.

5. The pH of blood tends to be slightly _____.

III. PLASMA IS THE FLUID COMPONENT OF BLOOD

Albumins	Gamma	Lymph
Clotting	Globulins	Serum
Fibrinogen	Liver	Water

1. Plasma consists mainly of _____.

Copyright © 1992 W. B. SAUNDERS COMPANY All rights reserved

2. When the proteins involved in clotting have been removed from plasma, the remaining liquid is called ——————————.

3. Plasma proteins may be divided into three groups, or fractions: the ——————————, ——————————, and ——————————.

4. One group of globulins, the —————————— globulins, serves as antibodies.

5. Fibrinogen and several other plasma proteins are important in the process of blood ——————————.

6. The gamma globulins are produced in the —————————— tissues, whereas the other plasma proteins are manufactured in the ——————————.

IV. RED BLOOD CELLS TRANSPORT OXYGEN

Anemia	Iron	Oxyhemoglobin
Arteries	Marrow	Red
Erythrocytes	Nucleus	Stem
Hemoglobin	Oxygen	Veins

1. An adult male has approximately 30 trillion —————————— blood cells circulating in his blood.

2. Red blood cells are also called ——————————.

3. Red blood cells are adapted for producing and packaging ——————————, the red pigment that transports oxygen.

4. A mature red blood cell lacks a(n) —————————— as well as most other organelles.

5. As blood circulates through the lungs, —————————— diffuses into the blood and into the red blood cells.

6. Oxygen combines weakly with hemoglobin to form ——————————.

7. Oxyhemoglobin is responsible for the color of the oxygen-rich blood that flows through the ——————————.

8. Hemoglobin that is not combined with oxygen is bluish in color and accounts for the darker appearance of blood that flows through the ——————————.

9. In children, red blood cells are produced in the red bone —————————— of almost all bones.

10. Bone marrow contains —————————— cells that multiply, giving rise to the blood cells.

11. Hemoglobin deficiency is a condition called ——————————.

12. The most common cause of anemia is a deficiency of —————————— in the diet.

Copyright © 1992 W. B. SAUNDERS COMPANY All rights reserved

V. WHITE BLOOD CELLS DEFEND THE BODY AGAINST DISEASE

Bacterial	Lymphocytes	Phagocytize (destroy)
Basophils	Macrophages	Stem
Eosinophils	Monocytes	Tissues
Leukocytes	Neutrophils	Viral

1. White blood cells are also called _____.

2. White blood cells develop from _____ cells in the red bone marrow, but some types complete their maturation elsewhere in the body.

3. Unlike red blood cells, many white blood cells leave the circulation and perform their functions in various _____.

4. As they move through the body, the white blood cells _____ bacteria, dead cells, and foreign matter.

5. The types of leukocytes that contain granules in their cytoplasm are _____, _____, and _____.

6. The two types of leukocytes that lack specific granules in their cytoplasm are _____ and _____.

7. Monocytes migrate into the connective tissues and develop into _____, the large scavenger cells of the body.

8. An elevated white blood cell count may indicate the presence of a(n) _____ infection.

9. _____ infections cause lowered white blood cell counts.

VI. PLATELETS FUNCTION IN BLOOD CLOTTING

Clot	Globulin	Serum
Clotting factors	Liver	Thrombin
Fibrin	Proteins	Thrombocytes

1. Platelets, also called _____, are tiny fragments of cytoplasm that become detached from certain cells in the bone marrow.

2. When you cut your finger, a complex series of chemical reactions produces tiny fibers that reinforce the platelets, forming a strong _____.

3. When tissue is damaged, a series of reactions takes place involving certain _____ in the blood. These are also called _____.

4. Prothrombin activator is an enzyme that catalyzes the conversion of prothrombin to its active form, _____.

5. Prothrombin is a _____ found in the plasma.

6. Prothrombin is manufactured in the _____ with the help of vitamin K.

Copyright © 1992 W. B. SAUNDERS COMPANY All rights reserved

7. Thrombin acts as an enzyme to convert the plasma protein fibrinogen to

    _____ .

8. Within a few minutes after clot formation, the clot begins to contract and squeeze out

    _____ .

Copyright © 1992 W. B. SAUNDERS COMPANY All rights reserved

Eleven

THE CIRCULATORY SYSTEM: THE HEART

OUTLINE

   I. The heart wall consists of three layers
  II. The heart has four chambers
 III. Valves prevent backflow of blood
 IV. The heart has its own blood vessels
  V. The conduction system consists of specialized cardiac muscle
 VI. The cardiac cycle includes contraction and relaxation phases
 VII. Heart sounds result from closing of the valves
VIII. Heart rate is regulated by the nervous system

LEARNING OBJECTIVES

After you have studied this chapter, you should be able to:
1. Describe the structure of the wall of the heart.
2. Identify the chambers of the heart and compare their functions.
3. Locate the atrioventricular and semilunar valves and compare their structures.
4. Identify the principal blood vessels that serve the heart wall.
5. Identify the components of the conduction system and trace the path of a muscle impulse through the heart.
6. Describe the events of the cardiac cycle, define systole and diastole, and correlate normal heart sounds with the events of the cardiac cycle.
7. Define cardiac output and explain how the heart is regulated.

Copyright © 1992 W. B. SAUNDERS COMPANY All rights reserved

NAME _____

STUDY QUESTIONS

Within each category, fill in the blanks with the correct response.

I. THE HEART WALL CONSISTS OF THREE LAYERS

Endocardium   Myocardium   Pericardium
Endothelial   Parietal   Wall
Muscle   Pericardial

1. The _____ of the heart is richly supplied with nerves, blood vessels, and lymph vessels.

2. From the inside out, the layers of the heart are the _____,

   _____, and _____ .

3. The endocardium consists of a smooth _____ lining resting on connective tissue.

4. The myocardium consists of the cardiac _____, which contracts to pump the blood.

5. The pericardium consists of two layers that are separated by a potential space, the

   _____ cavity.

6. The outer layer of the pericardium, the _____ pericardium, forms a strong sac for the heart and helps to anchor it within the thorax.

II. THE HEART HAS FOUR CHAMBERS

Atria   Interventricular   Septum
Auricle   Pulmonaray   Ventricles
Interatrial   Pump

1. The heart is a double _____ .

2. The right and left sides of the heart are completely separated by a wall, or

   _____ .

3. The _____ receive blood returning to the heart from the veins and act as reservoirs between contractions of the heart.

4. The _____ pump blood into the great arteries leaving the heart.

5. _____ arteries carry blood to the lungs, where gases are exchanged.

6. The wall between the atria is the _____ septum.

7. The wall between the ventricles is the _____ septum.

8. A small muscular pouch called a(n) _____ increases the surface area of each atrium.

Copyright © 1992 W. B. SAUNDERS COMPANY All rights reserved

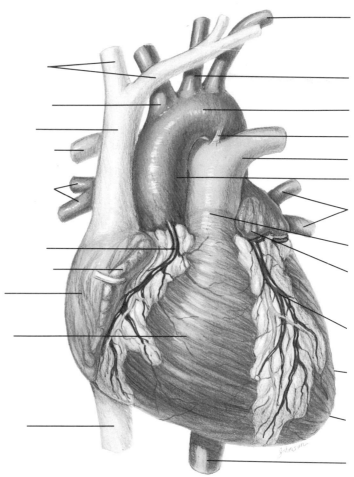

FIGURE 11–1   Fill in the correct labels.

III. VALVES PREVENT BACKFLOW OF BLOOD

Aortic	Cusps	Semilunar
Atrioventricular	Mitral	Tricuspid
Bicuspid	Pulmonary	

1. To prevent blood from flowing backward into the atrium, a(n) _____ valve guards the passageway between each atrium and ventricle.

2. The atrioventricular valve consists of flaps or _____ of fibrous tissues that project from the heart wall.

3. The atrioventricular valve between the right atrium and the right ventricle has three cusps and is called the _____ valve.

4. The left atrioventricular valve, which has only two cusps, is called the _____ valve.

5. The bicuspid valve is commonly called the _____ valve.

6. The _____ valves guard the exits from the ventricles.

Copyright © 1992 W. B. SAUNDERS COMPANY All rights reserved

7. The semilunar valve between the left ventricle and the aorta is called the
   _____ valve.

8. The semilunar valve between the right ventricle and the pulmonary artery is called
   the _____ valve.

IV. THE CONDUCTION SYSTEM CONSISTS OF SPECIALIZED CARDIAC MUSCLES

Atrioventricular	Conduction	Myocardium
Bundle	Intercalated	Sinoatrial

1. The heart has its own specialized _____ system and can beat independent of its nerve supply.

2. The _____ node is a small mass of specialized muscle in the posterior wall of the right atrium and is called the pacemaker of the heart.

3. The _____ node, located in the right atrium along the lower part of the septum, delays the transmission of an impulse to allow the atria to complete their contraction before the ventricles contract.

4. From the atrioventricular node, a muscle impulse spreads into specialized muscle fibers that form the atrioventricular _____.

5. The ends of the fibers of the atrioventricular bundle connect to fibers of ordinary cardiac muscle within the _____.

6. Cardiac muscle fibers are joined at their ends by dense bands called _____ discs.

V. THE CARDIAC CYCLE INCLUDES CONTRACTION AND RELAXATION PHASES

Arteries	Diastole	Veins
Atria	Semilunar	Ventricles
Cycle	Systole	

1. The period of contraction of the heart muscle is called _____.

2. The period of relaxation of the heart muscle is called _____.

3. Each cardiac _____ begins with a muscle impulse.

4. This impulse spreads from the sinoatrial node throughout the _____.

5. As the atria contract, the atrioventricular valves open, and blood is forced from the atria into the _____.

6. As blood is forced from the atria into the ventricles, the _____ valves close.

7. As the atria relax, they are filled with blood from the _____.

8. As the ventricles contract, blood is forced through the semilunar valves into the _____.

Copyright © 1992 W. B. SAUNDERS COMPANY All rights reserved

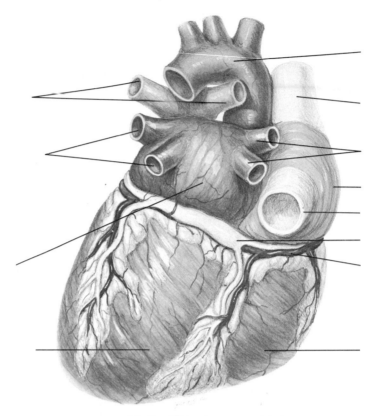

FIGURE 11-2 Fill in the correct labels.

VI. THE HEART IS REGULATED BY THE NERVOUS SYSTEM

Adrenal medulla	Cardiac	Vagus
Autonomic	Sympathetic	Venous

1. The _____ nervous system is responsible for changing the strength and rate of contraction of the heart.

2. Under conditions of stress, _____ nerves can increase the strength of contraction by as much as 100%.

3. Under calm conditions, the _____ nerve, a parasympathetic nerve, slows the heart.

4. During stress, epinephrine and norepinephrine released from the _____ increase the heartbeat.

5. The _____ output is the volume of blood pumped by one ventricle in 1 minute.

6. Cardiac output depends mainly on _____ return.

Copyright © 1992 W. B. SAUNDERS COMPANY All rights reserved

15. The heart's conduction system is made up of specialized _____ muscle.
    a. Conduction
    b. Cardiac
    c. Pulmonary
    d. Aortic
    e. None of the preceding

16. The _____ is a small mass of specialized muscle in the posterior wall of the right atrium.
    a. Atrioventricular node
    b. Sinoatrial node
    c. Sinoatrial bundle
    d. Atrioventricular bundle

Copyright © 1992 W. B. SAUNDERS COMPANY All rights reserved

Twelve

CIRCULATION OF BLOOD AND LYMPH

OUTLINE

  I. Three main types of blood vessels are arteries, capillaries, and veins
 II. The blood vessel wall consists of layers
III. Blood circulates through two circuits
   A. The pulmonary circulation carries blood to and from the lungs
   B. The systemic circulation carries blood to the tissues
      1. The aorta has four main regions
      2. Four arteries supply the brain
      3. The liver has an unusual circulation
 IV. Several factors influence blood flow
   A. The alternate expansion and recoil of an artery make up its pulse
   B. Blood pressure depends on blood flow and resistance to blood flow
   C. Pressure changes as blood flows through the systemic circulation
   D. Several mechanisms regulate blood pressure
   E. Blood pressure is expressed as systolic pressure over diastolic pressure
  V. The lymphatic system is a subsystem of the circulatory system
   A. Lymph nodes filter lymph
   B. Tonsils filter tissue fluid
   C. The spleen is the largest lymphatic organ
   D. The thymus gland plays a role in immune function

LEARNING OBJECTIVES

After you have studied this chapter, you should be able to:
1. Compare the structures and functions of arteries, capillaries, and veins.
2. Trace a drop of blood through the pulmonary and systemic circulations, listing the principal vessels and heart chambers through which it must pass on its journey from one part of the body to another. (For example, trace a drop of blood from the inferior vena cava to an organ such as the brain and then back to the heart.)
3. Identify the main divisions of the aorta and its principal branches.
4. Trace a drop of blood through the hepatic portal system.
5. Give the physiological basis for arterial pulse, and describe how pulse is measured.
6. Define blood pressure and give its relation to blood flow and resistance.
7. Compare blood pressures in the different types of blood vessels of the systemic circulation.
8. Describe the homeostatic mechanisms that regulate blood pressure and explain how blood pressure is measured.

118

Copyright © 1992 W. B. SAUNDERS COMPANY All rights reserved

NAME _____

STUDY QUESTIONS

Within each category, fill in the blanks with the correct response.

I. THREE MAIN TYPES OF BLOOD VESSELS ARE ARTERIES, CAPILLARIES, AND VEINS

Arteries	Oxygen	Veins
Arterioles	Pulmonary	Venules
Capillaries	Tissues	

1. Blood vessels carry blood into the _____ of the body.

2. _____ carry blood from the ventricles of the heart to each of the organs of the body.

3. All arteries except the _____ arteries carry blood rich in oxygen.

4. The smallest branches of an artery, called _____, are important in regulating blood pressure.

5. _____ permit materials to be exchanged between the blood and tissues.

6. Blood passes from the capillaries into _____, where it is conducted back toward the heart.

7. The smallest veins are called _____.

8. All veins except the pulmonary vein carry blood that is poor in _____.

II. THE BLOOD VESSEL WALL CONSISTS OF LAYERS

Arteries	Endothelium	Tunics
Collagen	Macrophages	Veins
Connective	Sinusoids	

1. The wall of an artery or vein has three layers, or _____.

2. The inner layer of a blood vessel consists of _____, which forms a smooth surface for the blood.

3. The middle layer of a blood vessel consists of _____ tissue and smooth muscle.

4. The outer layer of a blood vessel consists of connective tissue that is rich in elastic and _____ fibers.

5. In general, _____ have thinner walls than _____.

6. In the liver, spleen, and bone marrow, arterioles and venules are connected by capillary-like vessels called _____.

7. _____ lie along the outer walls of sinusoids. They reach into the vessels to remove worn-out blood cells and foreign matter from the circulation.

Copyright © 1992 W. B. SAUNDERS COMPANY All rights reserved

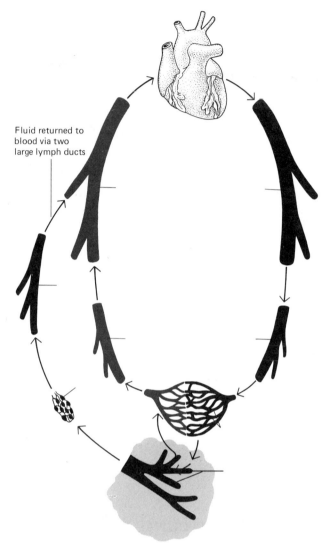

Fluid returned to
blood via two
large lymph ducts

FIGURE 12–1 Fill in the correct labels.

III. BLOOD CIRCULATES THROUGH TWO CIRCUITS

Abdominal   Hepatic portal vein   Superior vena cava
Aorta   Inferior vena cava   Systemic
Arch   Left   Thoracic
Ascending   Pulmonary
Atrium   Right

1. The _____ circulation connects the heart with the lungs.

2. The _____ circulation connects the heart with all of the organs and tissues.

3. The _____ ventricle pumps blood into the systemic circulation.

Copyright © 1992 W. B. SAUNDERS COMPANY All rights reserved

4. The _____ ventricle pumps blood into the pulmonary circulation.

5. From the pulmonary circulation, blood is returned to the left _____ .

6. The left ventricle pumps blood into the largest artery in the body, the

_____ .

7. The _____ receives blood from the upper portion of the body.

8. The _____ receives blood returning from below the level of the diaphragm.

9. The _____ aorta is the first part of the aorta and travels upward.

10. The aortic _____ curves from the ascending aorta and makes a U turn.

11. The _____ aorta descends from the aortic arch and passes through the thorax.

12. The _____ aorta is the region of the aorta below the diaphragm.

13. The _____ delivers blood from the organs of the digestive system to the liver.

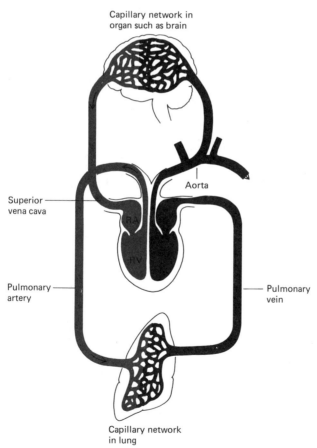

FIGURE 12–2   Fill in the correct labels.

Copyright © 1992 W. B. SAUNDERS COMPANY All rights reserved

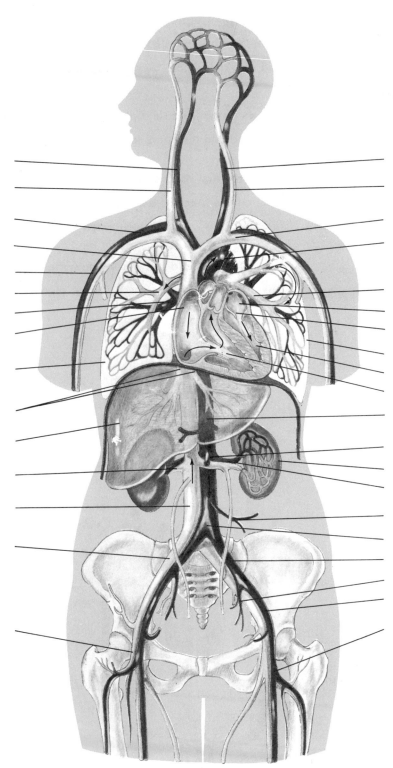

FIGURE 12–3 Fill in the correct labels.

Copyright © 1992 W. B. SAUNDERS COMPANY All rights reserved

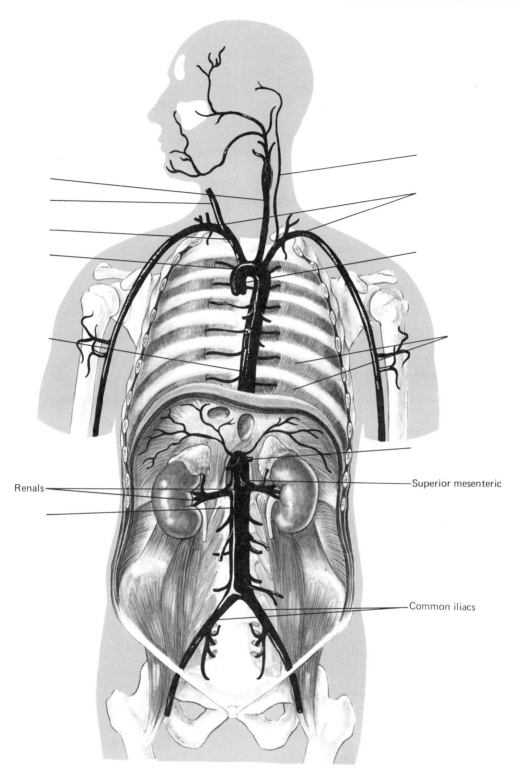

Renals

Superior mesenteric

Common iliacs

FIGURE 12–4   Fill in the correct labels.

Copyright © 1992 W. B. SAUNDERS COMPANY All rights reserved

IV. SEVERAL FACTORS INFLUENCE BLOOD FLOW

Angiotensins   Hypertension   Systole
Blood   Pulse   Systolic
Blood pressure   Pumping   Veins
Diastole   Radial   Volume
Heartbeats   Renin

1. The alternating expansion and recoil of an artery make up the arterial

    _____ .

2. The ability of the large arteries to expand and then snap back to their original

    diameter is important in maintaining a continuous flow of _____ .

3. The _____ artery in the wrist is the artery most frequently used to
    measure pulse.

4. The number of pulsations counted per minute indicates the number of

    _____ per minute.

5. _____ is the force exerted by the blood against the inner walls of the
    blood vessels.

6. The flow of blood depends directly on the _____ action of the heart.

7. The _____ of blood flowing through the body also affects blood
    pressure.

8. Very little pressure is needed to force the blood through the _____
    because they offer little resistance to blood flow.

9. Low blood pressure stimulates the kidneys to release the hormone

    _____ .

10. Renin converts a plasma protein to _____ , a group of hormones that
    act as powerful constrictors.

11. In arteries, blood pressure rises during _____ and falls during

    _____ .

12. _____ pressure may vary greatly with physical exertion and emotional
    stress.

13. High blood pressure is called _____ .

V. THE LYMPHATIC SYSTEM IS A SUBSYSTEM OF THE CIRCULATORY SYSTEM

Axillary   Duct   Plasma
Bacteria   Filter   Spleen
Blood   Lymph   Thoracic
Capillaries   Lymphatic   Thymus
Drainage   Lymphocytes   Tissue

1. The _____ system is a subsystem of the circulatory system.

Copyright © 1992 W. B. SAUNDERS COMPANY All rights reserved

2. The lymphatic system consists of the clear, watery lymph formed from

   _____ fluid.

3. The lymph circulation is a(n) _____ system.

4. The job of the lymphatic system is to collect excess tissue fluid and return

   it to the _____.

5. Lymph _____ conduct lymph to larger vessels called lymphatics.

6. Lymphatic vessels from throughout the body except the upper right quadrant drain

   into the _____ duct.

7. Lymph from the lymphatic vessels in the upper right quadrant of the body drains

   into the right lymphatic _____.

8. Lymph is continuously emptied into the blood, where it mixes with the

   _____.

9. _____ nodes are masses of lymph tissue surrounded by a connective
   tissue capsule.

10. The main functions of lymph nodes are to _____ the lymph and

    produce _____.

11. Lymph nodes are most numerous in the _____ and groin regions, and
    many are located in the thorax and abdomen.

12. When _____ are present, lymph nodes may increase in size and
    become tender.

13. The _____ is the largest organ of the lymphatic system.

14. The _____ gland is a pinkish-gray lymphatic organ located in the
    upper thorax anterior to the great vessels as they emerge from the heart and posterior
    to the sternum.

Copyright © 1992 W. B. SAUNDERS COMPANY All rights reserved

16. When cardiac output increases, _____.
    a. Blood flow increases
    b. Blood flow decreases
    c. Blood pressure increases
    d. Blood pressure decreases
    e. Both a and c

17. Blood can flow through the veins against gravity because _____.
    a. A system of valves prohibits backflow of blood
    b. Most blood is found in the arteries, not in the veins
    c. Veins are large and elastic and offer little resistance
    d. Both a and c
    e. All of the preceding

18. The principal function(s) of the lymphatic system is (are) _____.
    a. To collect and return tissue fluid to the blood
    b. To defend the body against disease by producing lymphocytes
    c. To absorb lipids from the intestine and transport them to the blood
    d. All of the preceding
    e. None of the preceding

19. Lymph is filtered by _____.
    a. Lymphatics
    b. Lymph nodes
    c. Lungs
    d. Liver
    e. None of the preceding

20. The _____ is the largest organ of the lymphatic system.
    a. Spleen
    b. Liver
    c. Lymph node
    d. Lymphatics
    e. None of the preceding

Copyright © 1992 W. B. SAUNDERS COMPANY All rights reserved

Thirteen

THE BODY'S DEFENSE MECHANISMS

OUTLINE

I. The body distinguishes self from nonself
II. Nonspecific defense mechanisms operate rapidly
III. Specific defense mechanisms include cell- and antibody-mediated immunities
    A. T cells are responsible for cell-mediated immunity
    B. B cells are responsible for antibody-mediated immunity
    C. There are five classes of antibodies
    D. Antibodies combine with antigens
    E. The thymus is important in immune function
    F. Active immunity develops after exposure to antigens
    G. Passive immunity is borrowed immunity

LEARNING OBJECTIVES

After you have studied this chapter, you should be able to:
1. Identify several nonspecific defense mechanisms.
2. Define the terms "antigen" and "antibody."
3. Describe cell-mediated immunity, including development of memory cells.
4. Describe antibody-mediated immunity, including the effects of antigen-antibody combination on pathogens both directly and through the complement system.
5. Describe the role of the thymus in immune mechanisms.
6. Contrast active with passive immunity and give examples of each.

Copyright © 1992 W. B. SAUNDERS COMPANY All rights reserved

NAME _____

STUDY QUESTIONS

Within each category, fill in the blanks with the correct response.

I. THE BODY DISTINGUISHES SELF FROM NONSELF

Antigen   Immunology   Proteins
Immune   Pathogens

1. The body has remarkable defense mechanisms that protect against

   _____, organisms that cause disease.

2. Specific defense mechanisms are collectively referred to as

   _____ responses.

3. _____ is the study of the body's specific defense mechanisms.

4. A single bacterium may have from 10 to more than 1000 distinct _____
   on its surface.

5. A substance capable of stimulating the body's defense mechanisms is called a(n)

   _____.

II. NONSPECIFIC DEFENSE MECHANISMS OPERATE RAPIDLY

Enzymes   Nonspecific   Sebum
Fever   Nose   Skin
Inflammation   Phagocytic   Sweat
Interferons   Phagocytosis
Mechanical   Respiratory

1. _____ defense mechanisms prevent pathogens from entering the body.

2. The _____ is the body's first line of defense against pathogens and
   other harmful substances.

3. The skin is a(n) _____ barrier that blocks the entry of pathogens.

4. _____ and _____, which are found on the surface of
   the skin, also contain chemicals that destroy certain types of bacteria.

5. Pathogens that enter the body with inhaled air may be filtered out by hairs in the

   _____ or trapped in the sticky mucous lining of the

   _____ passageway.

6. Bacteria that enter with food are usually destroyed by the acids and

   _____ of the stomach.

7. When pathogens invade tissues, _____ occurs.

8. The increased blood flow that occurs during inflammation brings

   _____ cells to the infected area.

Copyright © 1992 W. B. SAUNDERS COMPANY All rights reserved

9. _____ is a common clinical symptom of widespread inflammation.

10. One of the main functions of inflammation appears to be increased

_____ .

11. When infected by viruses or some types of bacteria or fungi, certain types of cells respond by secreting proteins called _____ .

III. SPECIFIC DEFENSE MECHANISMS INCLUDE CELL- AND ANTIBODY-MEDIATED IMMUNITIES

Antigen   Lymph   Passive
B   Lymphatic   Pathogen
Helper   Lymphocytes   T
IgG   Macrophages   Thymus
Immunization   Memory
Killer   Mitosis

1. Specific defense is the function of the _____ system.

2. The principal warriors in specific immune responses are the _____ .

3. Lymphocytes are stationed strategically in the _____ tissue throughout the body.

4. In cell-mediated immunity, _____ cells directly attack invading pathogens.

5. Cell-mediated immunity is the responsibility of T cells and _____ .

6. There are thousands of different types of T cells, each capable of responding to a specific type of _____ .

7. Once stimulated, T cells multiply by _____ , each giving rise to a sizable group of cloned cells identical to itself.

8. One type of specialized T cell, the _____ T cell, combines with antigens on the surface of an invading cell.

9. Some T cells remain behind in the lymph tissue and act as _____ cells.

10. _____ cells produce specific antibodies and send the antibodies out to perform their functions.

11. Certain T cells, the _____ T cells, play a role in the stimulation of B cells.

12. Normally, approximately 75% of the antibodies in the body belong in the _____ group.

13. The principal function of an antibody is to identify a(n) _____ as foreign.

14. The _____ helps T cells develop the ability to specialize.

15. Active immunity can be developed by _____ , that is, through injection of a vaccine.

16. In _____ immunity, an individual is given antibodies actively produced by other humans or animals.

Copyright © 1992 W. B. SAUNDERS COMPANY All rights reserved

Fourteen

THE RESPIRATORY SYSTEM

OUTLINE

I. The nasal cavities are lined with a mucous membrane
II. The pharynx is divided into three regions
III. The larynx is the voice box
IV. The trachea is the windpipe
V. The bronchi enter the lungs
VI. The air sacs are called alveoli
VII. The lungs provide a large surface area
VIII. Ventilation moves air into and out of the lungs
IX. Gas exchange occurs by diffusion
X. Gases are transported by the circulatory system
XI. Respiration is regulated by the brain
XII. The respiratory system defends itself against dirty air

LEARNING OBJECTIVES

After you have studied this chapter, you should be able to:
1. Trace a breath of air through the respiratory system from nose to alveoli.
2. Describe the structure of the lungs.
3. Define ventilation and summarize how breathing takes place.
4. Summarize the process of oxygen and carbon dioxide exchange in the lungs and in the tissues.
5. Outline how oxygen and carbon dioxide are transported in the blood.
6. Describe how respiration is regulated.

Copyright © 1992 W. B. SAUNDERS COMPANY All rights reserved

NAME _____

STUDY QUESTIONS

Within each category, fill in the blanks with the correct response.

I. INTRODUCTION

Bronchi	Nose	Respiratory
Lung	Pharynx	Trachea
Nasal	Respiration	

1. _____ supplies the cells of the body with oxygen and rids them of carbon dioxide.

2. The _____ system consists of the lungs and the tubes through which air reaches them.

3. A breath of air enters the body through the _____.

4. From the nose, air flows through the _____ cavities to the

   _____, through the larynx (voice box), and then into

   the _____ .

5. From the trachea, air enters the _____.

6. One bronchus enters each _____ .

II. THE NASAL CAVITIES CONTAIN A MUCOUS LINING

Filters	Pharynx	Smell
Moistens	Septum	Throat
Mucous	Sinuses	

1. Whether you breathe through your nose or your mouth, air finds its way into the

   _____ .

2. When breathing through the nose, the nose _____ and

   _____ the air and brings it to body temperature.

3. The nose contains the receptors for the sense of _____.

4. The nostrils open into the two nasal cavities, which are separated by a partition

   called the nasal _____ .

5. The septum and walls of the nose consist of bone covered with a

   _____ membrane.

6. Ciliated epithelial cells of the membrane push a steady stream of mucous, along with

   its trapped particles, toward the _____ .

7. Several _____ in the bones of the skull communicate with the nasal cavities through small channels.

Copyright © 1992 W. B. SAUNDERS COMPANY All rights reserved

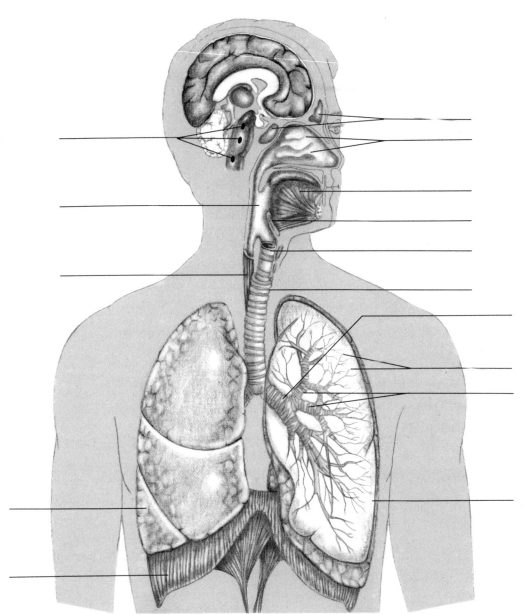

FIGURE 14–1 Fill in the correct labels.

III. THE PHARYNX IS DIVIDED INTO THREE REGIONS

Esophagus	Larynx	Oropharynx
Laryngopharynx	Nasopharynx	

1. Air enters the _____, the superior part of the pharynx.

2. From the nasopharynx, air moves down into the _____ behind the mouth.

3. From the oropharynx, air passes through the _____ and enters the

   _____ .

4. Behind the opening into the larynx, there is a second opening into the

   _____ .

Copyright © 1992 W. B. SAUNDERS COMPANY All rights reserved

IV. THE LARYNX IS THE VOICE BOX

Adam's apple	Epiglottis	Laryngitis
Cough	Glottis	Larynx

1. The _____, or voice box, contains the vocal cords.

2. The opening into the larynx is called the _____.

3. During swallowing, a flap of tissue called the _____ automatically closes off the larynx so that food cannot enter the lower airway.

4. When the epiglottis fails to close properly, foreign matter comes into contact with the sensitive larynx, which causes a(n) _____ reflex.

5. The wall of the larynx is supported by cartilage that protrudes from the midline of the neck and is sometimes called the _____.

6. Inflammation of the larynx, or _____, is often caused by a respiratory infection or by irritating substances such as cigarette smoke.

V. THE TRACHEA IS THE WINDPIPE

Cartilage	Lungs	Trachea
Esophagus	Mucous	
Larynx	Pharynx	

1. The _____, or windpipe, is located anterior to the esophagus.

2. The trachea extends from the _____ to the middle of the chest.

3. Like the larynx, the trachea is kept from collapsing by C-shaped rings of _____ in its wall.

4. The open parts of these rings of cartilage face posteriorly toward the _____.

5. The larynx, trachea, and bronchi are lined by a(n) _____ membrane that traps dirt and other foreign matter.

6. Ciliated cells in this lining continuously beat a stream of mucus upward to the _____, where it is swallowed.

7. This cilia-propelled mucus "elevator" keeps foreign material out of the _____.

VI. THE BRONCHI ENTER THE LUNGS/THE AIR SACS ARE CALLED ALVEOLI

Alveoli	Bronchus	Surfactant
Alveolus	Capillaries	Trachea
Bronchi	Lung	
Bronchioles	Respiratory	

1. The larynx is divided into the right and left _____.

Copyright © 1992 W. B. SAUNDERS COMPANY All rights reserved

2. One _____ enters each lung.

3. The structure of the main bronchi is similar to that of the _____.

4. Each bronchus branches repeatedly, giving rise to increasingly smaller bronchi and finally to very small _____ _____.

5. There are more than 1 million bronchioles in each _____.

6. The network of branching and rebranching in each lung is called the _____ tree.

7. Each bronchiole leads into a cluster of tiny air sacs, the _____.

8. The wall of a(n) _____ consists of a single layer of epithelial cells and elastic fibers that permit it to stretch and contract during breathing.

9. Each alveolus is surrounded by a network of _____ so that gases diffuse easily between the alveolus and blood.

10. Alveoli are coated with a thin film of _____, a substance that prevents them from collapsing.

VII. THE LUNGS PROVIDE A LARGE SURFACE AREA

Diaphragm	Pleural	Visceral pleura
Lobes	Pleural cavity	
Parietal pleura	Thoracic	

1. The lungs are large, paired, spongy organs that occupy the _____ cavity.

2. The right lung is divided into three _____.

3. Each lung is covered with a _____ membrane that forms a sac enclosing the lung and continues as the lining of the thoracic cavity.

4. The part of the pleural membrane that encloses the lung is the _____.

5. The part of the pleural membrane that lines the thoracic cavity is called the _____.

6. Between the pleural membranes is a potential space, the _____.

7. The floor of the thoracic cavity is a strong, dome-shaped muscle called the _____.

VIII. VENTILATION MOVES AIR INTO AND OUT OF THE LUNGS

Abdominal	Diaphragm	Inspiration
Breathing	Expiration	Pulmonary
Decreases	Increases	

1. _____ ventilation is the movement of air into and out of the lungs.

2. In general, we carry on pulmonary ventilation by _____.

3. The act of breathing in is called _____.

Copyright © 1992 W. B. SAUNDERS COMPANY All rights reserved

4. _____ occurs when the diaphragm and intercostal muscles relax.

5. During inspiration, the size of the chest cavity _____ .

6. During expiration, the size of the thoracic cavity _____ .

7. During forced expiration, several sets of muscles, including the

   _____ muscles, contract.

8. During inspiration, the _____ contracts and flattens, and the intercostal muscles contract.

IX. GAS EXCHANGE OCCURS BY DIFFUSION

Alveolus   Circulatory   Inspired
Breathing   Diffuses   Oxygen
Carbon dioxide   Expired

1. _____ delivers oxygen to the alveoli of the lungs.

2. The vital link between the alveoli and the body cells is the _____ system.

3. Each _____ serves as a depot from which oxygen is loaded into the blood of the pulmonary capillaries.

4. The alveoli contain a greater concentration of _____ than the blood entering the pulmonary capillaries.

5. _____ moves from the blood, where it is more concentrated, to the alveoli, where it is less concentrated.

6. Each gas _____ through the thin linings of the capillary and the alveolus.

7. _____ air contains 100 times more carbon dioxide than air

   _____ from the environment.

X. GASES ARE TRANSPORTED BY THE CIRCULATORY SYSTEM/RESPIRATION IS REGULATED BY THE BRAIN

Blood pressure   Hemoglobin   Phrenic
Carbon dioxide   Hyperventilate   Respiratory
CPR   Oxygen

1. When _____ diffuses into the blood, it enters the red blood cells and forms a weak chemical bond with hemoglobin, forming oxyhemoglobin.

2. Because the chemical bond linking oxygen with _____ is weak, this reaction is readily reversible.

3. The rate, depth, and rhythm of breathing are regulated by _____ centers in the medulla and pons.

4. During exercise, body tissues produce greater amounts of _____ .

5. Impulses from the medulla reach the diaphragm by way of the

   _____ nerves.

Copyright © 1992 W. B. SAUNDERS COMPANY All rights reserved

6. Underwater swimmers and divers not using scuba gear may voluntarily

   _____ to decrease the amount of carbon dioxide in their blood. This allows them to stay under water for a few extra moments before the urge to breathe becomes irresistible.

7. A certain concentration of carbon dioxide in the blood is necessary to maintain

   normal _____ .

8. _____ is a method for helping victims who have suffered respiratory and cardiac arrest.

XI. THE RESPIRATORY SYSTEM DEFENDS ITSELF AGAINST DIRTY AIR

Bronchial   Hair   Mucous lining
Carbon   Lung   Respiratory
Cilia   Lymph
Disease   Macrophages

1. The _____ system has a number of defense mechanisms that help protect the delicate lungs from damage.

2. The _____ in the nose and the _____ of the respiratory passageways help trap foreign particles in inspired air.

3. When we breathe dirty air, the _____ tubes narrow.

4. The smallest bronchioles and the alveoli are not equipped with mucus or cells with

   _____ .

5. Foreign particles that get through the respiratory defenses and find their way into the alveoli may remain there indefinitely, or they may be engulfed by _____ .

6. Macrophages may accumulate in the _____ tissue of the lungs.

7. Lung tissue of chronic smokers and those who work in dirty industrial environments contains large blackened areas where _____ particles have been deposited.

8. Continued insult to the respiratory system results in _____ .

9. Cigarette smoking is the main cause of _____ cancer.

Copyright © 1992 W. B. SAUNDERS COMPANY All rights reserved

Fifteen

THE DIGESTIVE SYSTEM

OUTLINE

I. The digestive system consists of the digestive tract and accessory organs
II. The digestive system processes food
III. The wall of the digestive tract has four layers
IV. The folds of the peritoneum support the digestive organs
V. The mouth ingests food
    A. The teeth break down food
    B. The salivary glands produce saliva
VI. The pharynx is important in swallowing
VII. The esophagus conducts food to the stomach
VIII. The stomach digests food
IX. Most digestion takes place in the small intestine
X. The pancreas secretes enzymes
XI. The liver secretes bile
XII. Digestion occurs as food moves through the digestive tract
    A. Glucose is the main product of carbohydrate digestion
    B. Bile digests fat
    C. Proteins are digested to amino acids
XIII. The intestinal villi absorb nutrients
XIV. The large intestine eliminates waste
XV. A balanced diet is necessary to maintain health

LEARNING OBJECTIVES

After you have studied this chapter, you should be able to:
1. List in sequence each structure through which a bite of food passes on its way through the digestive tract, and label a diagram of the digestive system.
2. Describe in general terms the following steps of food processing: ingestion, digestion, absorption, and elimination.
3. Describe the wall of the digestive tract, distinguish between the visceral and parietal peritoneums, and describe their major folds.
4. Describe the structures of the mouth, including the teeth, and give their functions.
5. Describe the structures and functions of the pharynx and esophagus.
6. Describe the structure of the stomach and its role in processing food.
7. Identify the three main regions of the small intestine and give the function of the small intestine.
8. Summarize the functions of the pancreas and liver.

Copyright © 1992 W. B. SAUNDERS COMPANY All rights reserved

9. Summarize carbohydrate, lipid, and protein digestion.
10. Describe the structure of an intestinal villus and explain the role of villi in absorption of nutrients.
11. Describe the structure and functions of the large intestine.
12. List the nutrients that make up a balanced diet, and summarize the functions of each nutrient.

Copyright © 1992 W. B. SAUNDERS COMPANY All rights reserved

NAME _____

STUDY QUESTIONS

Within each category, fill in the blanks with the correct response.

I. INTRODUCTION/THE DIGESTIVE SYSTEM CONSISTS OF THE DIGESTIVE TRACT AND ACCESSORY ORGANS

Alimentary	Liver	Nutrients
Anus	Metabolism	Pancreas
Gastrointestinal	Mouth	Salivary

1. _____ are the substances in food that we need as an energy source to run the machinery of the body.

2. The body uses nutrients as building blocks to make new cells and as ingredients to make the chemical compounds needed for _____.

3. The digestive tract, also called the _____ canal, is a tube approximately 4.4 m long.

4. The digestive tract extends from the _____, where food is taken in, to the _____, through which unused food is eliminated.

5. Below the diaphragm, the digestive tract is often referred to as the _____ tract.

6. Three types of accessory digestive glands are the _____ glands, _____, and _____. These are not part of the digestive tract, but they secrete digestive juices into it.

II. THE DIGESTIVE SYSTEM PROCESSES FOOD

Absorption	Elimination	Liver
Digestion	Ingestion	

1. _____ involves taking food into the mouth, chewing it, and then swallowing it.

2. _____ is the breakdown of food into smaller molecules.

3. _____ is the transfer of digested food through the wall of the intestine and into the circulatory system.

4. The circulatory system transports the food molecules, or nutrients, to the _____, where many are removed and stored.

5. _____ removes undigested and unabsorbed food from the body.

Copyright © 1992 W. B. SAUNDERS COMPANY All rights reserved

III. THE WALL OF THE DIGESTIVE TRACT HAS FOUR LAYERS

Connective	Mucosa	Peritonitis
Digestion	Parietal peritoneum	Submucosa
Digestive	Peristalsis	
Epithelial	Peritoneal cavity	

1. From the esophagus to the anus, the wall of the _____ tract consists of four layers.

2. The _____ is the lining of the digestive tract. It consists of _____ tissue resting on a layer of loose connective tissue.

3. In the stomach and small intestine, the mucosa is thrown into folds, which greatly increase its surface area for _____ and absorption.

4. Beneath the mucosa lies a layer of connective tissue called the _____, which is rich in blood vessels and nerves.

5. The third layer of the digestive system consists of muscle. This muscle contracts in a wavelike motion called _____, which pushes food through the digestive tract.

6. The outer layer of the wall of the digestive tract consists of _____ tissue.

7. The _____ is the sheet of connective tissue that lines the walls of the abdominal and pelvic cavities.

8. Between the visceral and parietal peritoneums is a potential space called the _____.

9. Inflammation of the peritoneum, called _____, can have very serious consequences, since infection can easily spread to adjoining organs.

IV. FOLDS OF THE PERITONEUM SUPPORT THE DIGESTIVE ORGANS

Greater omentum	Lesser omentum	Mesocolon
Intestine	Mesentery	Peritoneum

1. The _____, a large double fold of peritoneal tissue, extends from the parietal peritoneum and attaches to the small intestine.

2. The mesentery anchors the _____ to the posterior abdominal wall.

3. Other important folds of the _____ are the greater omentum, the lesser omentum, and the mesocolon.

4. The _____, also known as the "fatty apron," is a large double fold of peritoneum attached to the stomach and intestine.

5. The _____ suspends the stomach and duodenum from the liver.

6. The _____ is a fold of peritoneum that attaches the colon to the posterior abdominal wall.

Copyright © 1992 W. B. SAUNDERS COMPANY All rights reserved

V. THE MOUTH INGESTS FOOD

Alveolar	Oral cavity	Submandibular
Bolus	Parotid	Taste buds
Crown	Pulp cavity	Tongue
Dentin	Roots	
Enamel	Sublingual	

1. The mouth, or _____, ingests food and prepares it for digestion.

2. The flexible, muscular _____ on the floor of the mouth pushes the food around, which aids in chewing and swallowing.

3. _____ on the tongue enable us to taste foods as sweet, sour, salty, or bitter.

4. The teeth are rooted in sockets of the _____ processes.

5. Each tooth consists of a(n) _____, the portion above the gum, and one or more _____, the portion beneath the gum line.

6. Teeth are composed mainly of _____, a calcified connective tissue that imparts shape and rigidity to the tooth.

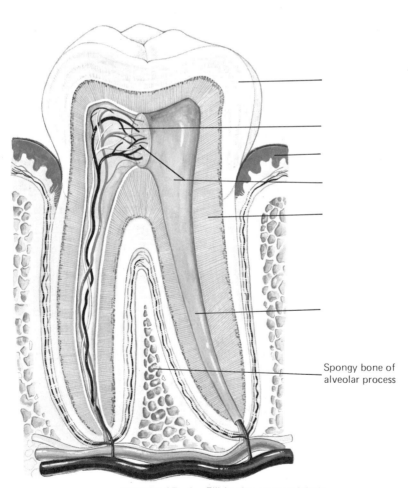

Spongy bone of alveolar process

FIGURE 15–1 Fill in the correct labels.

Copyright © 1992 W. B. SAUNDERS COMPANY All rights reserved

7. In the crown of the tooth, the dentin is protected by a tough covering of

_____.

8. The dentin encloses a(n) _____ filled with pulp, an extremely sensitive connective tissue containing blood vessels and nerves.

9. The _____ glands are the largest salivary glands.

10. The _____ glands lie below the jaw.

11. The _____ glands are under the tongue.

12. By moistening food, saliva helps the tongue convert a mouthful of food to a semisolid mass called a(n) _____.

VI. THE PHARYNX IS IMPORTANT IN SWALLOWING

Epiglottis	Nasopharynx	Soft palate
Esophagus	Oropharynx	Swallowing
Laryngopharynx	Pharynx	Tongue

1. _____ moves the bolus from the mouth through the pharynx and down the esophagus.

2. The _____, or throat, is a muscular tube approximately 12 cm long that serves as the hallway of the respiratory and digestive systems.

3. The three regions of the pharynx are the _____, posterior to the mouth; the _____, posterior to the nose; and the _____, which opens into the larynx and esophagus.

4. The oropharynx and the nasopharynx are partitioned by the _____.

5. In swallowing, the bolus is forced into the oropharynx by the _____.

6. Reflex movements of muscles in the wall of the pharynx propel the food into the

_____.

7. During swallowing, the opening to the larynx is closed by a small flap of tissue called the _____.

VII. THE ESOPHAGUS CONDUCTS FOOD TO THE STOMACH

Esophagus	Peristaltic	Sphincter
Heartburn	Pharynx	Stomach

1. The esophagus extends from the _____ through the thoracic cavity.

2. The esophagus passes through the diaphragm and empties into the _____.

3. The bolus is swept through the pharynx and into the esophagus by a wave of muscle contraction called a(n) _____ contraction.

4. At the lower end of the esophagus is a circular muscle called a(n)

_____ muscle.

Copyright © 1992 W. B. SAUNDERS COMPANY All rights reserved

5. The sphincter muscle prevents the highly acidic gastric juices from splashing up into the _____ .

6. Sometimes, gastric juices spurt up into the esophagus. The wall of the esophagus becomes irritated, and the resulting spasms cause _____ .

VIII. THE STOMACH DIGESTS FOOD

Chyme   Mucus   Pyloric
Contractions   Pepsin   Rugae
Glands   Peristalsis   Small intestine

1. When empty, the lining of the stomach has many folds, which are called

   _____ .

2. _____ of the stomach mix the food thoroughly.

3. The stomach mashes and churns food and moves it along by _____ .

4. The stomach is lined with simple epithelium that secretes large amounts of

   _____ .

5. Millions of gastric _____ in the wall of the stomach secrete hydrochloric acid and enzymes.

6. The main enzyme secreted in the gastric juice is _____ , which begins the digestion of proteins.

7. As food is digested over a period of 3 to 4 hours, it is converted into a soupy mixture

   called _____ .

8. The exit of the stomach is guarded by the _____ sphincter, a strong ring of muscle.

9. When the pyloric sphincter relaxes, chyme passes into the _____ .

IX. MOST DIGESTION TAKES PLACE IN THE SMALL INTESTINE

Absorbed   Duodenum   Pancreas
Digestion   Ileum   Small intestine
Duodenum   Jejunum   Villi

1. The _____ is a coiled tube more than 5 m long and 4 cm in diameter.

2. The first 22 cm or so of the small intestine is the _____ , which is curved like the letter C.

3. The portion of the small intestine that curves downward is called the

   _____ , and it extends for approximately 2 m; the last portion of the

   small intestine is called the _____ .

4. The lining of the small intestine has millions of tiny fingerlike projections called

   _____ .

Copyright © 1992 W. B. SAUNDERS COMPANY All rights reserved

5. The villi increase the surface area of the small intestine so that there is greater surface area for _____ and absorption of nutrients.

6. Most digestion takes place in the _____ rather than in the stomach.

7. The liver and _____ release digestive juices into the duodenum that act on the chyme.

8. The enzymes produced by the epithelial cells lining the duodenum complete the job of breaking down food molecules so that they can be _____.

X. THE PANCREAS SECRETES ENZYMES

Bile	Duodenum	Pancreas
Digestive	Exocrine	Pancreatic

1. The _____ is a long, large gland that lies in the abdomen inferior to the stomach.

2. The pancreas is both an endocrine and a(n) _____ gland.

3. The exocrine portion of the pancreas secretes _____ juice, which contains a number of _____ enzymes.

4. The pancreatic duct from the pancreas joins the _____ duct coming from the liver, forming a single duct that passes into the _____.

XI. THE LIVER SECRETES BILE

Common bile duct	Intestine	Lobe
Gallbladder	Liver	Portal
Hepatic	Liver cell	

1. The _____ is the largest and one of the most complex organs in the body.

2. A single _____ can carry on more than 500 separate metabolic activities.

3. The right _____ of the liver is larger than its left one and has three main parts.

4. Oxygen-rich blood is brought to the liver by the _____ arteries.

5. The liver also receives blood from the hepatic _____ vein.

6. The hepatic portal vein delivers nutrients absorbed from the _____.

7. Bile is stored in the pear-shaped _____.

8. The cystic duct from the gallbladder joins the duct from the liver to form the _____.

Copyright © 1992 W. B. SAUNDERS COMPANY All rights reserved

XII. DIGESTION OCCURS AS FOOD MOVES THROUGH THE DIGESTIVE TRACT

Amino acids   Duodenum   Peptide
Bile   Gastrin   Protein
Carbohydrate   Glucose   Stomach
Chyme   Mouth

1. Secretion of digestive juices is stimulated by hormones and _____.

2. The hormone _____, which is released by the stomach mucosa, stimulates the gastric glands to secrete.

3. Large carbohydrates such as starch and glycogen consist of long chains of _____ molecules.

4. Starch digestion begins in the _____.

5. Glucose is the main product of _____ digestion.

6. Digestion of fat takes place mainly in the _____.

7. _____ emulsifies fat by a detergent action that breaks down large fat droplets into smaller droplets.

8. Proteins consist of smaller molecules called _____.

9. Amino acid subunits are linked together by chemical bonds called _____ bonds.

10. Protein digestion begins in the _____.

11. The products of _____ digestion are free amino acids.

XIII. THE INTESTINAL VILLI ABSORB NUTRIENTS

Absorbed   Lacteal   Villi
Fatty acids   Liver   Villus

1. After food has been digested, the nutrients are absorbed by the intestinal _____.

2. Within each _____ is a network of capillaries that branch from an arteriole and empty into a venule.

3. A central lymph vessel called a _____ is also inside the villus.

4. Amino acids and simple sugars are _____ into the blood.

5. Amino acids and simple sugars are transported directly to the _____ by the hepatic portal vein.

6. _____ are absorbed into the lacteals.

Copyright © 1992 W. B. SAUNDERS COMPANY All rights reserved

XIV. THE LARGE INTESTINE ELIMINATES WASTES

Anal canal	Chyme	Rectum
Appendicitis	Colon	Sigmoid colon
Ascending colon	Ileocecal	Transverse colon
Cecum	Peristaltic	Vermiform appendix

1. After the _____ has passed through the stomach and small intestine, it consists mainly of water and indigestible wastes such as cellulose.

2. The small intestine is separated from the large intestine by the _____ valve.

3. When a _____ contraction brings chyme toward it, the ileocecal valve opens, allowing the chyme to enter the large intestine.

4. The first 7 cm of the large intestine is a pouch called the _____.

5. The _____, a worm-shaped blind tube, hangs down from the end of the cecum.

6. Inflammation of the appendix, called _____, can lead to peritonitis and other complications if not diagnosed and treated promptly.

7. From the cecum to the rectum, the large intestine is called the _____.

8. The _____ extends from the cecum straight up to the lower edge of the liver.

9. As the ascending colon turns horizontally, it becomes the _____.

10. On the left side of the abdomen, the descending colon turns downward and forms the S-shaped _____, which empties into the short _____.

11. The last 4 cm of the rectum are called the _____.

XV. A BALANCED DIET IS NECESSARY TO MAINTAIN HEALTH

Carbohydrates	Minerals	Vitamins
Lipids	Proteins	Water

1. _____ is one of the main components of the body and is used by the body to transport materials.

2. _____ are inorganic nutrients ingested in the form of salts dissolved in food and water.

3. _____ are organic compounds required for certain reactions to take place. Many of these serve as coenzymes, compounds that work with enzymes to regulate chemical reactions.

4. _____ are ingested mainly as starch or cellulose. These are digested to glucose or other simple sugars, which can be absorbed into the blood.

5. Most of the _____ we ingest are fats or cholesterols.

6. _____ are digested into their component amino acids. Proteins are the essential building blocks of cells, and many serve as enzymes.

Copyright © 1992 W. B. SAUNDERS COMPANY All rights reserved

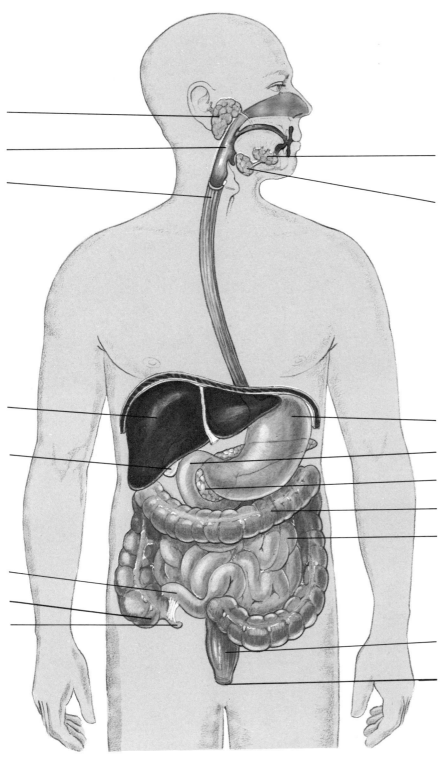

FIGURE 15–2 Fill in the correct labels.

Copyright © 1992 W. B. SAUNDERS COMPANY All rights reserved

NAME _____

Crossword Puzzle for Chapters 14 and 15

ACROSS

1 Windpipe
5 During digestion, many carbohydrates are broken down into these components.
6 Number of bronchi
8 Large salivary gland
9 Produced by liver
10 Terminal part of small intestine
12 Large muscle that functions in breathing
14 Contains vocal cords
15 Large organ that lies inferior to diaphragm

DOWN

1 Consists of crown and roots
2 Opening for elimination of feces
3 Cavity in which lungs are located
4 Large intestine from cecum to rectum
5 Stores bile
7 Digests fat
11 Throat
13 Absorb nutrients

Copyright © 1992 W. B. SAUNDERS COMPANY All rights reserved

Sixteen

THE URINARY SYSTEM

OUTLINE

  I. Metabolic wastes and their secretion
 II. Organs of the urinary system
III. The kidneys
    A. The nephron
    B. Urine formation
       1. Glomerular filtration
       2. Tubular reabsorption and secretion
       3. Urine composition
       4. Urine volume regulation
    C. Overview of kidney function
 IV. Urine transport and storage
  V. Urination

LEARNING OBJECTIVES

After you have studied this chapter, you should be able to:
1. Identify the principal metabolic waste products and the organs that excrete them.
2. Label a diagram of the structures of the urinary system and give the function of each structure.
3. Describe the structures of a nephron and give the functions of Bowman's capsule, glomerulus, renal tubule, collecting duct, afferent arteriole, and efferent arteriole. (Be able to label a diagram of a nephron.)
4. Trace a drop of filtrate from glomerulus to urethra, listing in sequence each structure through which the drop passes.
5. Describe the process of urine formation and give the composition of urine.
6. Summarize the regulation of urine volume, including the role of ADH.
7. Summarize the functions of the kidney in maintaining homcostasis.
8. Describe the process of urination.

Copyright © 1992 W. B. SAUNDERS COMPANY All rights reserved

11. Water is absorbed from the digestive tract into the _____.

12. Excess _____ is removed from the blood by the kidneys.

13. Coffee, tea, and alcoholic beverages contain chemicals called _____ that increase urine volume.

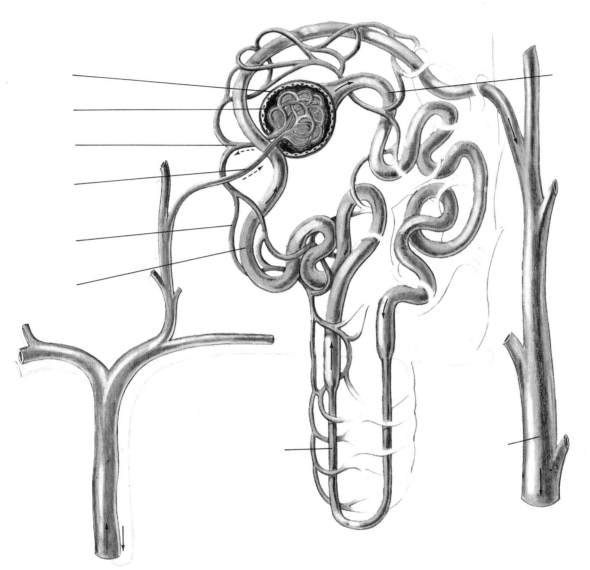

FIGURE 16–2 Fill in the correct labels.

IV. URINE IS TRANSPORTED BY DUCTS AND STORED IN THE BLADDER

Bladder	Prostate	Urinary bladder
Penis	Ureters	Vagina
Peristaltic contractions	Urethra	

1. Urine passes from the kidneys through the paired _____.

2. Urine is forced through the ureter by _____.

Copyright © 1992 W. B. SAUNDERS COMPANY All rights reserved

3. The _____ is a temporary storage sac for urine.

4. When urine leaves the bladder, it flows through the _____, a duct leading to the outside of the body.

5. In the male, the urethra is long and passes through the _____ gland and the _____.

6. In the female, the urethra is short and is found just above the opening into the

    _____.

7. _____ infections are more common in females than males because the long male urethra is a barrier to bacterial invasion.

Copyright © 1992 W. B. SAUNDERS COMPANY All rights reserved

Seventeen

REGULATION OF FLUIDS AND ELECTROLYTES

OUTLINE

  I. The body has two main fluid compartments
 II. Fluid intake must equal fluid output
    A. The hypothalamus regulates fluid intake
    B. The hypothalamus regulates fluid output
III. Electrolyte balance is affected by fluid balance
    A. Sodium is the major extracellular cation
    B. Potassium is the major intracellular cation
    C. Other major electrolytes include calcium, phosphate, chloride, and magnesium

LEARNING OBJECTIVES

After you have studied this chapter, you should be able to:
1. Identify the fluid compartments of the body.
2. Summarize the principal routes of fluid input and output.
3. Describe how fluid input and output are regulated.
4. Define electrolyte balance and identify the functions of six major electrolytes.
5. Describe the mechanisms responsible for sodium and potassium homeostasis.

Copyright © 1992 W. B. SAUNDERS COMPANY All rights reserved

NAME _____

STUDY QUESTIONS

Within each category, fill in the blanks with the correct response.

I. INTRODUCTION

Body fluids	Glucose	Urea
Electrolytes	Homeostasis	Water

1. Fluid balance is critical to _____ .

2. When we drink a large volume of water, we increase the amount of

   _____ .

3. The term "body fluid" refers to the _____ in the body and the substances dissolved in it.

4. Among the most important components of body fluid are _____ .

5. Examples of nonelectrolytes in the body fluid are _____ and

   _____ .

II. THE BODY HAS TWO MAIN FLUID COMPARTMENTS

Arteriole	Intracellular	Potassium ion
Blood pressure	Lymphatic	Sodium ion
Electrolytes	Osmotic concentration	Tissue fluid
Extracellular	Osmotic pressure	Water

1. The human body is approximately 60% _____ by weight.

2. Water and _____ are distributed in certain regions, or compartments.

3. About two thirds of body fluid is found in the _____ compartment.

4. The remaining third of body fluid is found in the _____ compartment.

5. The movement of fluid from one compartment to another depends on

   _____ and _____ .

6. Blood pressure forces fluid out of the blood at the _____ ends of capillaries.

7. When it leaves the blood, this fluid is called _____ .

8. Excess tissue fluid returns to the blood at the venous ends of capillaries because of

   _____ .

9. Excess tissue fluid is also returned to the blood by way of the _____ system.

10. _____ concentration is much higher in the extracellular fluid than in the intracellular fluid.

11. _____ concentration is much higher in the intracellular fluid than in the extracellular fluid.

Copyright © 1992 W. B. SAUNDERS COMPANY All rights reserved

III. FLUID INTAKE MUST EQUAL FLUID OUTPUT

Blood
Catabolic

Dehydration
Digestive tract

Kidneys
Water

1. We ingest approximately 2500 ml of _____ _____ each day in the foods we eat and liquids we drink.

2. Water is absorbed from the digestive tract into the _____.

3. Water is also produced during _____ processes.

4. Most fluid is discharged by the _____.

5. Fluid is also lost through the skin, lungs, and _____.

6. When fluid output is greater than fluid input, _____ occurs.

IV. THE HYPOTHALAMUS REGULATES FLUID INTAKE

Fluids
Hypothalamus

Osmotic
Saliva

Thirst center

1. Fluid intake is regulated by the _____.

2. Dehydration raises the _____ pressure of the blood.

3. Increased osmotic pressure stimulates the _____ in the hypothalamus.

4. Stimulating the thirst center results in the sensation of thirst and the desire to drink

_____.

5. Dehydration leads to a decrease in _____ secretion, which results in dryness in the mouth.

V. THE HYPOTHALAMUS REGULATES FLUID OUTPUT

ADH
Dehydrated
Hypothalamus

Kidneys
Pituitary gland
Plasma

Urine

1. The _____ are primarily responsible for fluid output.

2. Fluid output is regulated by _____.

3. ADH is produced by the hypothalamus and secreted by the posterior lobe of the

_____.

4. ADH regulates the volume of _____.

5. When the body begins to get _____, ADH secretion increases.

6. When the volume of water in the body decreases, the _____ becomes saltier.

7. Special receptors in the _____ signal the posterior pituitary to release more ADH.

Copyright © 1992 W. B. SAUNDERS COMPANY All rights reserved

VI. ELECTROLYTE BALANCE IS AFFECTED BY FLUID BALANCE

Anions   Ions   Water
Cations   Potassium
Electrolyte balance   Sodium

1. When the total amount of the various electrolytes taken into the body equals the amount lost, the body is in _____.

2. Electrolytes produce positively and negatively charged _____.

3. Negatively charged ions are called _____.

4. Positively charged ions are called _____.

5. Approximately 90% of extracellular cations are _____ ions.

6. Sodium concentration is adjusted mainly by regulating the amount of _____ in the body.

7. Most of the cations in the intracellular fluid are _____ ions.

Copyright © 1992 W. B. SAUNDERS COMPANY All rights reserved

NAME _____

Crossword Puzzle for Chapters 16 and 17

ACROSS

1 Duct that conducts urine from kidney to urinary bladder
3 Compound that forms ions
4 Indentation on medial border of kidney
7 Type of capillary that conducts blood to nephron
9 Microscopic unit in kidney consisting of renal corpuscle and renal tubule
11 Positively charged ion
12 Network of capillaries surrounded by Bowman's capsule

DOWN

1 Conducts urine from urinary bladder
2 Main component of body fluid
5 Nitrogen waste
6 Filtrate of flows from the proximal convoluted tubule to the loop of _____.
8 Main extracellular cation
10 Outer tissue of kidney
11 Outer connective tissue covering kidney

Copyright © 1992 W. B. SAUNDERS COMPANY All rights reserved

Eighteen

REPRODUCTION

OUTLINE

   I. The male produces sperm and delivers them into the female
   - A. The testes produce sperm
   - B. The conducting tubes transport sperm
   - C. The accessory glands produce semen
   - D. The penis delivers sperm into the female reproductive tract
   - E. Male hormones regulate male sexuality and reproduction
   II. The female produces ova and incubates the egg
   - A. The ovaries produces ova and hormones
   - B. The uterine tubes transport ova
   - C. The uterus incubates the embryo
   - D. The vagina functions in intercourse, menstruation, and birth
   - E. The vulva are the external genital structures
   - F. The breasts contain the mammary glands
   - G. Hormones regulate female reproduction
   - H. The menstrual cycle prepares the body for pregnancy
   - I. Menopause is marked by decline in ovary function
   III. Fertilization is the fusion of sperm and ovum
   IV. The zygote gives rise to the new individual
   - A. The embryo develops in the wall of the uterus
   - B. Prenatal development requires about 266 days
   - C. The birth process has three stages
   - D. Multiple births may be fraternal or identical
   V. The human life cycle extends from fertilization to death

LEARNING OBJECTIVES

After you have studied this chapter, you should be able to:
   1. Label a diagram of the male reproductive system and describe the functions of each structure.
   2. Trace the passage of sperm from the tubules in the testes through the conducting tubes, describing changes that may occur along the way.
   3. Describe the actions of the male gonadotropic hormones and of testosterone.
   4. Label diagrams of internal and external female reproductive organs and describe their structures and functions.
   5. Trace the development of an ovum and its passage through the female reproductive system.

178

Copyright © 1992 W. B. SAUNDERS COMPANY All rights reserved

6. Describe the principal events of the menstrual cycle and summarize the interactions of hormones that regulate the cycle.
7. Describe the process of fertilization and identify the time of the menstrual cycle at which sexual intercourse is most likely to result in pregnancy.
8. Summarize the course of human development from fertilization to birth.
9. Describe the functions of the amnion and the placenta.
10. Identify the three stages of the birth process.
11. List the stages of human development from fertilization to death.

Copyright © 1992 W. B. SAUNDERS COMPANY All rights reserved

NAME _____

STUDY QUESTIONS

Within each category, fill in the blanks with the correct response.

I. THE MALE PRODUCES SPERM AND DELIVERS THEM INTO THE FEMALE

Circumcision	Prepuce	Spermatic cord
Ejaculation	Prostate	Testes
Epididymis	Scrotum	Testosterone
Erect	Semen	Vas deferens
Glans	Shaft	
Penis	Sperm	

1. The male's function in reproduction is to produce _____ cells and deliver them into the female reproductive tract.

2. In the adult male, millions of sperm cells are manufactured each day within the paired male gonads, the _____.

3. The testes develop in the abdominal cavity of the male embryo. About 2 months before birth, they descend into the _____, a skin-covered sac suspended from the groin.

4. From the tubules inside the testes, sperm pass into a large, coiled tube, the

   _____.

5. The epididymis empties into a straight tube, the _____, or sperm duct.

6. The vas deferens passes from the scrotum through the inguinal canal as part of the

   _____.

7. _____ is a thick, whitish fluid consisting of sperm cells suspended in secretions of the accessory glands.

8. The single _____ gland surrounds the urethra as the urethra emerges from the urinary bladder.

9. The _____ is the male copulatory organ.

10. The penis consists of a long _____ that enlarges to form an expanded tip, the _____.

11. Part of the loose-fitting skin of the penis folds down and covers the proximal portion of the glans, forming a cuff called the _____, or foreskin.

12. The foreskin is removed during _____.

13. When the male becomes excited, spongy tissue in the penis fills with blood, and the penis becomes _____.

14. When the level of sexual excitement reaches a peak, _____ occurs.

15. _____ is the hormone responsible for the development of both primary and secondary sex characteristics in the male.

Copyright © 1992 W. B. SAUNDERS COMPANY All rights reserved

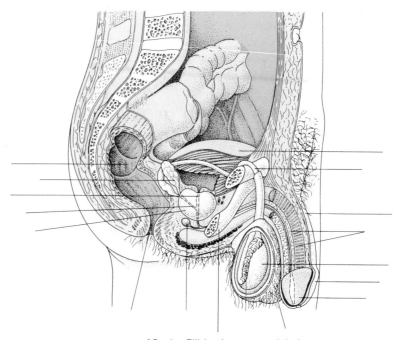

FIGURE 18-1   Fill in the correct labels.

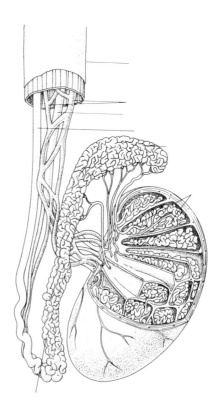

FIGURE 18-2   Fill in the correct labels.

Copyright © 1992 W. B. SAUNDERS COMPANY All rights reserved

II. THE FEMALE PRODUCES OVA AND INCUBATES THE EMBRYO

Breasts	Menstrual cycle	Progesterone
Cervix	Mons pubis	Uterine
Estrogen	Ova	Uterus
Lactation	Ovaries	Vagina
Menopause	Ovulation	Vulva

1. The paired _____ are the female gonads.

2. The ovaries produce _____ and the female sex hormones,

   _____ and _____ .

3. During _____ , the ovum is ejected through the wall of the ovary and into the pelvic cavity.

4. During ovulation, the mature ovum is drawn into the _____ tube.

5. Each month during a woman's reproductive life, the _____ , or womb, prepares for possible pregnancy.

6. The _____ functions as the sexual organ that receives the penis during sexual intercourse.

7. The vagina surrounds the end of the _____ .

8. The term "_____" refers to the external female genital structures.

9. The _____ is a mound of fatty tissue that covers the pubic symphysis.

10. The mammary glands are located within the _____ .

11. The function of the breasts is _____ .

12. The _____ stimulates production of an ovum each month and prepares the uterus for pregnancy.

13. At about age 50, a woman enters _____ .

III. FERTILIZATION IS THE FUSION OF SPERM AND OVUM

Ejaculation	Ovulation	Uterus
Follicle	Ovum	Zygote

1. When sperm are released in the vagina, some find their way into the

   _____ and uterine tubes.

2. Large numbers of sperm are necessary to penetrate the _____ cells surrounding the ovum (egg).

3. As soon as one sperm penetrates the _____ , no other sperm is able to get into the ovum.

4. Sperm and ovum fuse to form a fertilized egg, or _____ .

5. After _____ , sperm remain viable for only about 48 hours.

6. The ovum remains fertile for approximately 24 hours after _____ .

Copyright © 1992 W. B. SAUNDERS COMPANY All rights reserved

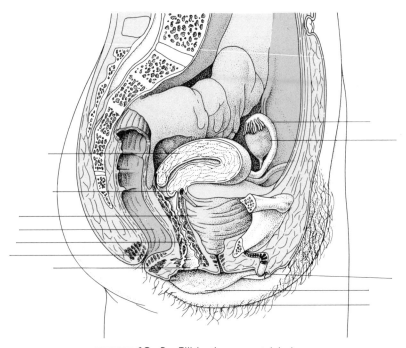

FIGURE 18–3 Fill in the correct labels.

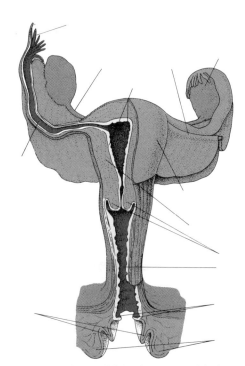

FIGURE 18–4 Fill in the correct labels.

Copyright © 1992 W. B. SAUNDERS COMPANY All rights reserved

IV. THE ZYGOTE GIVES RISE TO THE NEW INDIVIDUAL

Amnion   Fetus   Uterus
Embryo   Placenta   Zygote
Fetal   Umbilical cord

1. The _____ contains all of the genetic information to produce a
   complete individual.

2. The zygote divides to form a(n) _____ that is composed of two cells.

3. On about the seventh day of the development, the embryo begins to implant itself in

   the wall of the _____ .

4. Several _____ membranes develop around the embryo. They help
   protect, nourish, and support the developing embryo.

5. The _____ is a membrane that forms a sac around the embryo.

6. The _____ is the organ of exchange between the mother and the embryo.

7. The stalk of tissue that connects the embryo with the placenta is the

   _____ .

8. After the second month, the embryo is called a(n) _____ .

Copyright © 1992 W. B. SAUNDERS COMPANY All rights reserved

NAME _____

POST TEST

Fill in each blank with the letter that represents the correct response.

1. Reproduction involves several processes, including _____.
   a. Formation of specialized sex cells
   b. Preparation of the female body for pregnancy
   c. Sexual intercourse
   d. Both a and c
   e. All of the preceding

2. Male reproductive structures include the _____.
   a. Testes
   b. Penis
   c. Scrotum
   d. Both a and b
   e. All of the preceding

3. The sequence of the path that sperm pass through is _____.
   a. Tubules in the testis, epididymis, vas deferens, urethra, and ejaculatory duct
   b. Tubules in the testis, vas deferens, epididymis, ejaculatory duct, and urethra
   c. Tubules in the testis, epididymis, vas deferens, ejaculatory duct, and urethra
   d. Epididymis, vas deferens, urethra, and ejaculatory duct
   e. None of the preceding

4. The _____ is the male copulatory organ.
   a. Penis
   b. Testis
   c. Scrotum
   d. Both b and c
   e. All of the preceding

5. The _____ is removed during circumcision.
   a. Shaft
   b. Prepuce
   c. Glans
   d. All of the preceding
   e. None of the preceding

6. Primary sex characteristics in the male include _____.
   a. Growth of the penis and scrotum
   b. Growth and activity of internal reproductive structures
   c. Muscle development
   d. Both a and b
   e. All of the preceding

7. Secondary sex characteristics in the male include _____.
   a. Muscle development
   b. Deepening of the voice
   c. Growth of facial, pubic, and underarm hair
   d. Both b and c
   e. All of the preceding

Copyright © 1992 W. B. SAUNDERS COMPANY All rights reserved

8. The organs of the female reproductive system include the _____
   a. Ovaries
   b. Uterine tubes
   c. Uterus
   d. Vagina
   e. All of the preceding

9. The _____ are the female gonads.
   a. Uterine tubes
   b. Ovaries
   c. Breasts
   d. All of the preceding
   e. None of the preceding

10. In a normal pregnancy, the fetus develops in the _____.
    a. Ovaries
    b. Scrotum
    c. Uterus
    d. Uterine tubes
    e. None of the preceding

11. The term "vulva" refers to the external female genital structures, which include the

    _____.
    a. Labia
    b. Clitoris
    c. Mons pubis
    d. Both a and c
    e. All of the preceding

12. The mammary glands are located within the _____.
    a. Breasts
    b. Vagina
    c. Cervix
    d. Uterus
    e. None of the preceding

13. Although there is variation, a "typical" menstrual cycle is _____ days long.
    a. 3
    b. 7
    c. 28
    d. 45
    e. 365

14. During the menstrual cycle, ovulation occurs approximately _____ days before the next cycle begins.
    a. 7
    b. 14
    c. 28
    d. 36
    e. 128

15. By the fourth week of development, many organs in the fetus have begun to develop, including the _____.
    a. Brain
    b. Hands
    c. Feet
    d. Spinal cord
    e. a and d only

Copyright © 1992 W. B. SAUNDERS COMPANY All rights reserved

NAME _____

Crossword Puzzle for Chapter 18

ACROSS

3 Organ that produces sperm
5 Fertilized egg
8 Female sexual organ that receives the penis
9 Female gonads
10 Gland that surrounds the male urethra
11 Delivers sperm into the vagina
13 Eggs
14 Male sex hormone
16 Portion of the uterus that projects into the vagina
17 Stage at which menstrual cycle stops

DOWN

1 Organ through which the fetus receives nutrients
2 Monthly sloughing of the uterine lining
4 Membrane that partly blocks the entrance to the vagina
6 Release of ovum from the ovary
7 Female sex hormones
12 Glands located in the breasts
15 Sac that contains the testes

Copyright © 1992 W. B. SAUNDERS COMPANY All rights reserved

ANSWER KEY

CHAPTER ONE

STUDY QUESTIONS

SECTION I. INTRODUCTION TO ANATOMY AND PHYSIOLOGY

1. Anatomy
2. Physiology
3. Adapted
4. Stomach
5. Shape, structure

SECTION II. LEVELS OF ORGANIZATION IN THE BODY

1. Atoms
2. Matter
3. Ion
4. Molecules
5. Water
6. Cells
7. Microscope
8. Organelles
9. Nucleus
10. Tissue
11. Functions
12. Muscle, nervous, epithelial, connective
13. Organs
14. Body
15. Organism

SECTION III. LEVELS OF BIOLOGICAL ORGANIZATION IN THE BODY

1. Body
2. Metabolism
3. Catabolism, anabolism
4. Breaking down
5. Cellular respiration
6. ATP
7. Nutrients, oxygen
8. Building, synthetic

Copyright © 1992 W. B. SAUNDERS COMPANY All rights reserved

SECTION IV. HOMEOSTATIC MECHANISMS

1. Regulated
2. Homeostasis
3. Stressor
4. Feedback system
5. Negative feedback
6. Negative
7. Positive feedback
8. Positive

SECTION V. THE BASIC PLAN OF THE BODY

1. Mirror, bilateral symmetry
2. Cranium, vertebral column
3. Anatomical
4. Superior
5. Inferior
6. Closer
7. Cephalic
8. Caudad
9. Anterior, ventral
10. Posterior, dorsal
11. Axis
12. Medial
13. Midline
14. Lateral
15. Proximal
16. Distal
17. Superficial
18. Deep

SECTION VI. THE THREE MAIN PLANES OF THE BODY

1. Body planes
2. Sagittal plane
3. Midsagittal plane
4. Transverse plane
5. Frontal plane

SECTION VII. REGIONS OF THE BODY AND BODY CAVITIES

1. Axial
2. Appendicular
3. Torso
4. Body cavities
5. Viscera
6. Dorsal, ventral
7. Cranial cavity, spinal canal
8. Thoracic, abdominopelvic
9. Diaphragm
10. Pleural sacs, mediastinum

Copyright © 1992 W. B. SAUNDERS COMPANY All rights reserved

11. Pericardial
12. Abdominal cavity
13. Pelvic cavity

POST TEST

1. a
2. b
3. b
4. c
5. d
6. a
7. b
8. a
9. b
10. a
11. b
12. b
13. c
14. c
15. b
16. c
17. b

CHAPTER TWO

STUDY QUESTIONS

SECTION I. THE CELL CONTAINS SPECIALIZED ORGANELLES

1. Building blocks
2. Functions
3. Ovum
4. Cytoplasm
5. Amino acids
6. Organelles
7. Plasma membrane
8. Regulates
9. Endoplasmic reticulum
10. Smooth, rough
11. Ribosomes
12. Proteins
13. Golgi complex
14. Lysosomes
15. Enzymes
16. Mitochondria

Copyright © 1992 W. B. SAUNDERS COMPANY All rights reserved

17. Cellular respiration
18. Cilia
19. Nucleus
20. Chromosomes
21. Nucleolus

SECTION II. MATERIALS MOVE THROUGH THE PLASMA MEMBRANE

1. Permeable
2. Diffusion
3. Osmosis
4. Filtration
5. Active transport
6. Phagocytosis

SECTION III. CELLS DIVIDE, FORMING GENETICALLY IDENTICAL CELLS

1. Mitosis
2. Chromosomes
3. Five
4. Interphase
5. Prophase
6. Metaphase
7. Anaphase
8. Telophase
9. Two

SECTION IV. TISSUES ARE THE FABRIC OF THE BODY

1. Intercellular substance
2. Epithelial
3. Connective
4. Nervous
5. Muscle
6. Protection
7. Simple, stratified
8. Gland
9. Endocrine, exocrine
10. Join together
11. Organ
12. Contract
13. Involuntary
14. Smooth
15. Neurons, glial
16. Cell body
17. Dendrites, axon

SECTION V. MEMBRANES COVER OR LINE BODY SURFACES

1. Membranes
2. Synovial

Copyright © 1992 W. B. SAUNDERS COMPANY All rights reserved

3. Mucous
4. Serous
5. Parietal
6. Visceral

POST TEST

1. c
2. d
3. a
4. b
5. b
6. a
7. c
8. d
9. a
10. e
11. b
12. e
13. d
14. c
15. b
16. a
17. b
18. c
19. d
20. e
21. a
22. e
23. e

CHAPTER THREE

STUDY QUESTIONS

SECTION I. THE SKIN FUNCTIONS AS A PROTECTIVE BARRIER

1. Integumentary system
2. Sweat
3. Fluid
4. Vitamin D
5. Sensory receptors

SECTION II. THE SKIN CONSISTS OF THE EPIDERMIS AND DERMIS

1. Epidermis
2. Dermis

Copyright © 1992 W. B. SAUNDERS COMPANY All rights reserved

3. Subcutaneous
4. Epithelial
5. Outer
6. Deepest
7. Keratin
8. Connective
9. Collagen
10. Hair follicles, glands
11. Upper
12. Fingerprints
13. Superficial fascia
14. Adipose

SECTION III. SWEAT GLANDS AND SEBACEOUS GLANDS ARE FOUND IN THE SKIN

1. Sweat gland
2. Body temperature
3. Heat, raise
4. Evaporation
5. Water
6. Sebaceous glands
7. Ducts
8. Sebum
9. Acne
10. Pimple

SECTION IV. HAIR AND NAILS ARE APPENDAGES OF THE SKIN

1. Protective
2. Palms, soles
3. Shaft
4. Root
5. Follicle
6. Capillaries
7. Keratin
8. Dead
9. Smooth
10. Contract
11. Nails
12. Pink

SECTION V. MELANIN HELPS DETERMINE SKIN COLOR

1. Lowest
2. Melanin
3. Color
4. Albino
5. Sun
6. Absorbs
7. Darker
8. Ultraviolet
9. Sunburned
10. Cancer

Copyright © 1992 W. B. SAUNDERS COMPANY All rights reserved

POST TEST

1. c
2. b
3. e
4. c
5. a
6. b
7. c
8. b
9. b
10. e
11. b
12. a
13. b
14. a
15. c
16. a
17. c
18. d
19. e
20. e

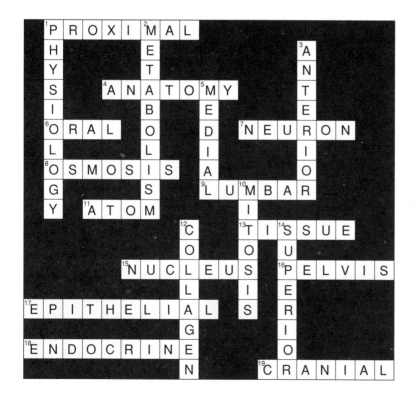

Copyright © 1992 W. B. SAUNDERS COMPANY All rights reserved

CHAPTER FOUR

STUDY QUESTIONS

SECTION I. FUNCTIONS OF THE SKELETAL SYSTEM

1. Supports, protects
2. Bones
3. Tendons
4. Ligaments
5. Marrow
6. Interaction

SECTION II. THE LONG BONE

1. Diaphysis
2. Epiphysis
3. Metaphysis
4. Epiphyseal
5. Yellow
6. Endosteum
7. Periosteum
8. Bone
9. Hyaline

SECTION III. TYPES OF BONE TISSUE

1. Compact, spongy
2. Dense
3. Spindle, osteons
4. Osteocytes, lacunae
5. Haversian canals
6. Canaliculi
7. Epiphyses
8. Bone marrow

SECTION IV. BONE DEVELOPMENT AND DIVIDING THE SKELETON

1. Ossification
2. Fetal
3. Endochondral
4. Intramembranous
5. Osteoblasts
6. Apatite
7. Lacunae
8. Osteocytes
9. Bones
10. Tissue, marrow
11. Osteoclasts, resorption
12. Enzymes
13. Finished
14. Two
15. Axial
16. Appendicular

Copyright © 1992 W. B. SAUNDERS COMPANY All rights reserved

SECTION V. THE SKULL

1. Cranial, facial
2. Eight
3. 14
4. Middle
5. Sutures
6. Sagittal suture
7. Frontal
8. Lambdoidal suture
9. Parietal
10. Anterior
11. Soft spots
12. Sinuses
13. Paranasal
14. Sinusitis

SECTION VI. THE VERTEBRAL COLUMN

1. Spine
2. 24
3. Sacrum, coccyx
4. Cervical, thoracic, five, fused, coccygeal
5. Intervertebral discs
6. Centrum
7. Spinous

SECTION VII. THE THORACIC CAGE

1. Rib
2. Pectoral
3. Sternum, thoracic, 12

SECTION VIII. THE PECTORAL AND PELVIC GIRDLES

1. Pectoral girdle
2. Clavicle
3. Sternum
4. Pelvic girdle
5. Trunk
6. Coxal, sacrum, coccyx
7. Pelvis
8. Inlet
9. Pubic

SECTION IX. THE UPPER AND LOWER LIMBS

1. 30
2. Humerus
3. Ulna, radius
4. Carpal
5. Metacarpals

Copyright © 1992 W. B. SAUNDERS COMPANY All rights reserved

6. Phalanges
7. Femur
8. Patella
9. Tibia, fibula
10. Tarsal
11. Metatarsals

SECTION X. THE JOINTS

1. Articulation
2. Joints
3. Movement
4. Synthroses, connective
5. Amphiarthroses, cartilage
6. Synovial
7. Diarthroses
8. Synovial fluid
9. Bursae
10. Bursitis

POST TEST

1. d
2. c
3. b
4. c
5. a
6. b
7. c
8. e
9. d
10. b
11. d
12. e
13. a
14. b
15. c
16. c
17. d
18. b
19. d
20. c
21. a
22. a
23. a
24. c

Copyright © 1992 W. B. SAUNDERS COMPANY All rights reserved

CHAPTER FIVE

STUDY QUESTIONS

SECTION I. INTRODUCTION

1. Muscles
2. Skeletal, smooth, cardiac
3. Voluntary

SECTION II. EACH SKELETAL MUSCLE IS AN ORGAN

1. Fibers
2. Epimysium
3. Fascicles
4. Perimysium
5. Endomysium
6. Tendons

SECTION III. MUSCLE FIBERS ARE SPECIALIZED FOR CONTRACTION

1. Nuclei
2. Transverse
3. Filaments
4. Myosin
5. Actin
6. Contractile

SECTION IV. MUSCLE CONTRACTION OCCURS WHEN ACTIN AND MYOSIN FILAMENTS SLIDE PAST EACH OTHER

1. Bones
2. Fibers
3. Actin, myosin
4. Motor
5. Impulses
6. Neuromuscular
7. Acetylcholine
8. Receptors
9. Action potential
10. Cholinesterase
11. Calcium
12. Muscle

SECTION V. ATP PROVIDES ENERGY FOR MUSCLE CONTRACTION

1. ATP
2. Creatine phosphate
3. Fuel
4. Glucose

Copyright © 1992 W. B. SAUNDERS COMPANY All rights reserved

5. Muscle fatigue
6. Lactic acid
7. Oxygen debt

SECTION VI. MUSCLE TONE IS A STATE OF PARTIAL CONTRACTION

1. Muscle tone
2. Unconscious
3. Upright
4. Motor nerve
5. Isotonic
6. Isometric

SECTION VII. MUSCLES WORK ANTAGONISTICALLY TO ONE ANOTHER

1. Tendons
2. Articulates
3. Origin
4. Insertion
5. Agonist
6. Antagonist
7. Synergists
8. Fixators

SECTION VIII. FUNCTIONS OF MUSCLES

1. Masseter
2. Trapezius
3. Rectus abdominis
4. Diaphragm
5. Pectoralis
6. Biceps
7. Gluteus maximus
8. Gastrocnemius

POST TEST

1. e
2. a
3. b
4. c
5. d
6. b
7. b
8. c
9. c
10. a
11. b
12. a
13. d

Copyright © 1992 W. B. SAUNDERS COMPANY All rights reserved

14. c
15. a
16. b
17. a
18. b
19. c
20. b
21. a

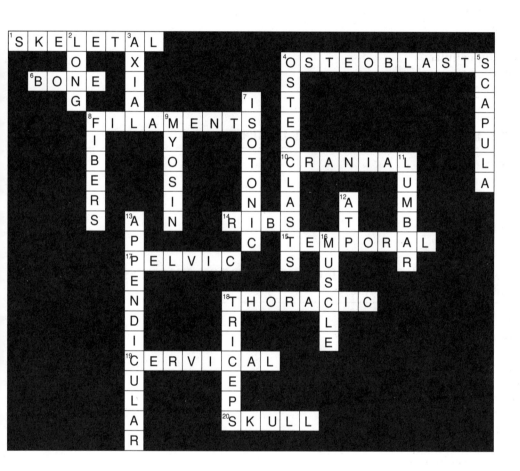

CHAPTER SIX

STUDY QUESTIONS

SECTION I. THE NERVOUS SYSTEM CONSISTS OF THE CENTRAL NERVOUS SYSTEM AND THE PERIPHERAL NERVOUS SYSTEM

1. Homeostasis
2. Central, peripheral

Copyright © 1992 W. B. SAUNDERS COMPANY All rights reserved

3. Brain
4. Sense
5. Cranial, spinal
6. Somatic, autonomic
7. Afferent
8. Efferent

SECTION II. NEURONS AND GILAL CELLS ARE THE CELLS OF THE NERVOUS SYSTEM

1. Glial
2. Neurons
3. Fibers
4. Dendrites
5. Axon
6. Neurotransmitters
7. Myelin, cellular
8. Neurilemma
9. Impulses

SECTION III. BUNDLES OF AXONS MAKE UP NERVES

1. Nerve
2. Axons, myelin
3. Ganglion
4. Tracts
5. Nuclei

SECTION IV. NEURAL FUNCTION INCLUDES RECEPTION, TRANSMISSION, INTEGRATION, AND RESPONSE

1. (3)
   (1)
   (4)
   (2)
   (5)
2. Neurons
3. Dendrites
4. Synapse
5. Synaptic
6. Neurotransmitters

SECTION V. THE HUMAN BRAIN IS THE MOST COMPLEX MECHANISM KNOWN

1. Brain
2. Stroke (cerebrovascular accident)
3. Brainstem
4. Ventricles
5. Oblongata
6. Cardiac, vasomotor
7. Pons
8. Medulla

Copyright © 1992 W. B. SAUNDERS COMPANY All rights reserved

9. Midbrain
10. Thalamus
11. Cerebellum
12. Cerebrum
13. Cortex
14. Ventricles

SECTION VI. THE SPINAL CORD TRANSMITS INFORMATION TO AND FROM THE BRAIN

1. Spinal
2. Vertebral
3. Fissures
4. Ascending
5. Descending

SECTION VII. THE CENTRAL NERVOUS SYSTEM IS WELL PROTECTED

1. Meninges
2. Dura mater
3. Arachnoid
4. Pia mater
5. Sinuses
6. Meningitis
7. Encephalitis
8. Cerebrospinal
9. Spinal
10. Reflex

POST TEST

1. b
2. c
3. d
4. e
5. a
6. b
7. c
8. d
9. e
10. d
11. e
12. e

Copyright © 1992 W. B. SAUNDERS COMPANY All rights reserved

CHAPTER SEVEN

STUDY QUESTIONS

SECTION I. THE SOMATIC SYSTEM RESPONDS TO CHANGES IN THE OUTSIDE WORLD

1. Sense
2. Somatic
3. Autonomic
4. Cranial, spinal
5. Dorsal, ventral
6. Cervical

SECTION II. THE AUTONOMIC SYSTEM MAINTAINS INTERNAL BALANCE

1. Autonomic
2. Sympathetic, parasympathetic
3. Efferent
4. Conserves
5. Brain
6. Vagus
7. Pelvic

POST TEST

1. a
2. c
3. b
4. e
5. d
6. c
7. b
8. a
9. b
10. a

CHAPTER EIGHT

STUDY QUESTIONS

SECTION I. INTRODUCTION

1. Stimulus
2. Sight, hearing, taste, smell, touch
3. Receptors

Copyright © 1992 W. B. SAUNDERS COMPANY All rights reserved

SECTION II. THE EYE CONTAINS VISUAL RECEPTORS

1. Orbit
2. Fat
3. Lashes, lids
4. Reflex
5. Blinking
6. Lacrimal
7. Ducts
8. Extrinsic
9. Sclera
10. Cornea
11. Conjunctiva
12. Iris
13. Pupil
14. Sensory
15. Cones
16. Rods
17. Fovea
18. Optic
19. Optic disc

SECTION III. THE EAR FUNCTIONS IN HEARING AND EQUILIBRIUM

1. Pinna
2. Middle
3. Cerumen
4. Tympanic
5. Temporal
6. Labyrinth
7. Perilymph
8. Membranous
9. Endolymph
10. Cochlea
11. Corti
12. Vestibule, semicircular

SECTION IV. SMELL IS SENSED BY RECEPTORS IN THE NASAL CAVITY

1. Olfactory
2. Nerve
3. Smell

SECTION V. TASTE IS SENSED BY THE TASTE BUDS

1. Tongue
2. Decreases
3. Solution
4. Tip
5. Sides
6. Back
7. Appetite

Copyright © 1992 W. B. SAUNDERS COMPANY All rights reserved

SECTION VI. THE GENERAL SENSES ARE WIDESPREAD THROUGHOUT THE BODY

1. Tactile
2. Lips, mouth, anus
3. Pain, tissue
4. Pain
5. Referred
6. Radiate
7. Acupuncture
8. Proprioceptors

POST TEST

1. c
2. a
3. e
4. d
5. c
6. b
7. d
8. a
9. d
10. c
11. d
12. a
13. d
14. b
15. c
16. a
17. b
18. c

CHAPTER NINE

STUDY QUESTIONS

SECTION I. INTRODUCTION

1. Endocrine
2. Hormones
3. Target
4. Endocrinology

SECTION II. HORMONES ACT ON TARGET CELLS

1. Proteins
2. Receptor

Copyright © 1992 W. B. SAUNDERS COMPANY All rights reserved

3. Messenger
4. Plasma
5. Prostaglandins
6. Local

SECTION III. ENDOCRINE GLANDS ARE REGULATED BY FEEDBACK CONTROL

1. Feedback
2. Parathyroid
3. Inhibited
4. Negative
5. Positive
6. Secretion
7. Hyposecretion
8. Hypersecretion

SECTION IV. THE HYPOTHALAMUS AND PITUITARY GLAND WORK TOGETHER CLOSELY

1. Hypothalamus
2. Hormones
3. Pituitary
4. Posterior
5. Oxytocin
6. ADH
7. Anterior
8. Protein

SECTION V. THE THYROID GLAND IS LOCATED IN THE NECK

1. Thyroid
2. Isthmus
3. Thyroxine
4. Metabolic
5. TSH
6. Hypothyroidism
7. Goiter
8. Iodine

SECTION VI. PARATHYROID GLANDS ARE LOCATED ON THE THYROID

1. Parathyroid
2. PTH
3. Calcium
4. Calcitonin
5. Decreased
6. Increased

Copyright © 1992 W. B. SAUNDERS COMPANY All rights reserved

SECTION VII. THE ISLETS OF LANGERHANS ARE THE ENDOCRINE PORTION OF THE PANCREAS

1. Pancreas
2. Langerhans
3. Beta
4. Alpha
5. Insulin
6. Glycogen
7. Glucagon
8. Sugar
9. Mellitus
10. Decrease
11. Noninsulin
12. Hyperglycemic

SECTION VIII. THE ADRENAL GLANDS FUNCTION IN METABOLISM AND STRESS

1. Adrenal
2. Medulla, cortex
3. Sympathetic
4. Epinephrine, norepinephrine
5. Adrenalin
6. Steroids
7. Glucocorticoids
8. Cortisol
9. Glucose
10. Mineralocorticoids
11. Aldosterone
12. Sex
13. ACTH

SECTION IX. STRESS THREATENS HOMEOSTASIS

1. Homeostasis
2. Stressors
3. Stress
4. Alarm
5. Resistance
6. Exhaustion

POST TEST

1. c
2. e
3. a
4. b
5. c
6. d
7. e
8. e

Copyright © 1992 W. B. SAUNDERS COMPANY All rights reserved

9. a
10. b
11. c
12. a
13. b
14. d
15. c
16. a
17. b
18. d

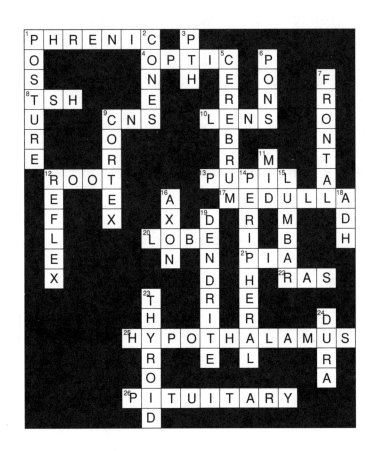

CHAPTER TEN

STUDY QUESTIONS

SECTION I. INTRODUCTION

1. Circulatory
2. Blood

Copyright © 1992 W. B. SAUNDERS COMPANY All rights reserved

3. Cardiovascular, lymphatic
4. Heart
5. Fluid

SECTION II. BLOOD CONSISTS OF CELLS AND PLATELETS SUSPENDED IN PLASMA

1. Red, white
2. Platelets
3. Plasma
4. 5.6, 6
5. Alkaline

SECTION III. PLASMA IS THE FLUID COMPONENT OF BLOOD

1. Water
2. Serum
3. Albumins, globulins, fibrinogen
4. Gamma
5. Clotting
6. Lymph, liver

SECTION IV. RED BLOOD CELLS TRANSPORT OXYGEN

1. Red
2. Erythrocytes
3. Hemoglobin
4. Nucleus
5. Oxygen
6. Oxyhemoglobin
7. Arteries
8. Veins
9. Marrow
10. Stem
11. Anemia
12. Iron

SECTION V. WHITE BLOOD CELLS DEFEND THE BODY AGAINST DISEASE

1. Leukocytes
2. Stem
3. Tissues
4. Phagocytize (destroy)
5. Neutrophils, basophils, eosinophils
6. Lymphocytes, monocytes
7. Macrophages
8. Bacterial
9. Viral

SECTION VI. PLATELETS FUNCTION IN BLOOD CLOTTING

1. Thrombocytes
2. Clot

Copyright © 1992 W. B. SAUNDERS COMPANY All rights reserved

3. Proteins, clotting factors
4. Thrombin
5. Globulin
6. Liver
7. Fibrin
8. Serum

POST TEST

1. d
2. e
3. e
4. e
5. a
6. b
7. a
8. d
9. c
10. d
11. b
12. c
13. a
14. c
15. d

CHAPTER ELEVEN

STUDY QUESTIONS

SECTION I. THE HEART WALL CONSISTS OF THREE LAYERS

1. Wall
2. Endocardium, myocardium, pericardium
3. Endothelial
4. Muscle
5. Pericardial
6. Parietal

SECTION II. THE HEART HAS FOUR CHAMBERS

1. Pump
2. Septum
3. Atria
4. Ventricles
5. Pulmonary
6. Interatrial

Copyright © 1992 W. B. SAUNDERS COMPANY All rights reserved

7. Interventricular
8. Auricle

SECTION III. VALVES PREVENT BACKFLOW OF BLOOD

1. Atrioventricular
2. Cusps
3. Tricuspid
4. Bicuspid
5. Mitral
6. Semilunar
7. Aortic
8. Pulmonary

SECTION IV. THE CONDUCTION SYSTEM CONSISTS OF SPECIALIZED CARDIAC MUSCLE

1. Conduction
2. Sinoatrial
3. Atrioventricular
4. Bundle
5. Myocardium
6. Intercalated

SECTION V. THE CARDIAC CYCLE INCLUDES CONTRACTION AND RELAXATION PHASES

1. Systole
2. Diastole
3. Cycle
4. Atria
5. Ventricles
6. Semilunar
7. Veins
8. Arteries

SECTION VI. THE HEART IS REGULATED BY THE NERVOUS SYSTEM

1. Autonomic
2. Sympathetic
3. Vagus
4. Adrenal medulla
5. Cardiac
6. Venous

POST TEST

1. d
2. a
3. c
4. a
5. b

Copyright © 1992 W. B. SAUNDERS COMPANY All rights reserved

6. b
7. a
8. c
9. a
10. d
11. c
12. b
13. c
14. a
15. b
16. b

CHAPTER TWELVE

STUDY QUESTIONS

SECTION I. THREE MAIN TYPES OF BLOOD VESSELS ARE ARTERIES, CAPILLARIES, AND VEINS

1. Tissues
2. Arteries
3. Pulmonary
4. Arterioles
5. Capillaries
6. Veins
7. Venules
8. Oxygen

SECTION II. THE BLOOD VESSEL WALL CONSISTS OF LAYERS

1. Tunics
2. Endothelium
3. Connective
4. Collagen
5. Veins, arteries
6. Sinusoids
7. Macrophages

SECTION III. BLOOD CIRCULATES THROUGH TWO CIRCUITS

1. Pulmonary
2. Systemic
3. Left
4. Right
5. Atrium
6. Aorta
7. Superior vena cava

Copyright © 1992 W. B. SAUNDERS COMPANY All rights reserved

8. Inferior vena cava
9. Ascending
10. Arch
11. Thoracic
12. Abdominal
13. Hepatic portal vein

SECTION IV. SEVERAL FACTORS INFLUENCE BLOOD FLOW

1. Pulse
2. Blood
3. Radial
4. Heartbeats
5. Blood pressure
6. Pumping
7. Volume
8. Veins
9. Renin
10. Angiotensins
11. Systole, diastole
12. Systolic
13. Hypertension

SECTION V. THE LYMPHATIC SYSTEM IS A SUBSYSTEM OF THE CIRCULATORY SYSTEM

1. Lymphatic
2. Tissue
3. Drainage
4. Blood
5. Capillaries
6. Thoracic
7. Duct
8. Plasma
9. Lymph
10. Filter, lymphocytes
11. Axillary
12. Bacteria
13. Spleen
14. Thymus

POST TEST

1. c
2. b
3. d
4. a
5. b
6. c
7. d
8. b
9. e

Copyright © 1992 W. B. SAUNDERS COMPANY All rights reserved

10. d
11. b
12. d
13. a
14. d
15. e
16. e
17. d
18. d
19. b
20. a

CHAPTER THIRTEEN

STUDY QUESTIONS

SECTION I. THE BODY DISTINGUISHES SELF FROM NONSELF

1. Pathogens
2. Immune
3. Immunology
4. Proteins
5. Antigen

SECTION II. NONSPECIFIC DEFENSE MECHANISMS OPERATE RAPIDLY

1. Nonspecific
2. Skin
3. Mechanical
4. Sweat, sebum
5. Nose, respiratory
6. Enzymes
7. Inflammation
8. Phagocytic
9. Fever
10. Phagocytosis
11. Interferons

SECTION III. SPECIFIC DEFENSE MECHANISMS INCLUDE CELL- AND ANTIBODY-MEDIATED IMMUNITIES

1. Lymphatic
2. Lymphocytes
3. Lymph
4. T
5. Macrophages

Copyright © 1992 W. B. SAUNDERS COMPANY All rights reserved

6. Antigen
7. Mitosis
8. Killer
9. Memory
10. B
11. Helper
12. IgG
13. Pathogen
14. Thymus
15. Immunization
16. Passive

POST TEST

1. a
2. b
3. c
4. b
5. d
6. c
7. d
8. a
9. c
10. c
11. b
12. d
13. d
14. a
15. a
16. b
17. b

Copyright © 1992 W. B. SAUNDERS COMPANY All rights reserved

CHAPTER FOURTEEN

STUDY QUESTIONS

SECTION I. INTRODUCTION

1. Respiration
2. Respiratory
3. Nose
4. Nasal, pharynx, trachea
5. Bronchi
6. Lung

SECTION II. THE NASAL CAVITIES CONTAIN A MUCOUS LINING

1. Pharynx
2. Filters, moistens
3. Smell
4. Septum
5. Mucous

Copyright © 1992 W. B. SAUNDERS COMPANY All rights reserved

6. Throat
7. Sinuses

SECTION III. THE PHARYNX IS DIVIDED INTO THREE REGIONS

1. Nasopharynx
2. Oropharynx
3. Laryngopharynx, larynx
4. Esophagus

SECTION IV. THE LARYNX IS THE VOICE BOX

1. Larynx
2. Glottis
3. Epiglottis
4. Cough
5. Adam's apple
6. Laryngitis

SECTION V. THE TRACHEA IS THE WINDPIPE

1. Trachea
2. Larynx
3. Cartilage
4. Esophagus
5. Mucous
6. Pharynx
7. Lungs

SECTION VI. THE BRONCHI ENTER THE LUNGS/THE AIR SACS ARE CALLED ALVEOLI

1. Bronchi
2. Bronchus
3. Trachea
4. Bronchioles
5. Lung
6. Respiratory
7. Alveoli
8. Alveolus
9. Capillaries
10. Surfactant

SECTION VII. THE LUNGS PROVIDE A LARGE SURFACE AREA

1. Thoracic
2. Lobes
3. Pleural
4. Visceral pleura
5. Parietal pleura
6. Pleural cavity
7. Diaphragm

Copyright © 1992 W. B. SAUNDERS COMPANY All rights reserved

SECTION VIII. VENTILATION MOVES AIR INTO AND OUT OF THE LUNGS

1. Pulmonary
2. Breathing
3. Inspiration
4. Expiration
5. Increases
6. Decreases
7. Abdominal
8. Diaphragm

SECTION IX. GAS EXCHANGE OCCURS BY DIFFUSION

1. Breathing
2. Circulatory
3. Alveolus
4. Oxygen
5. Carbon dioxide
6. Diffuses
7. Expired, inspired

SECTION X. GASES ARE TRANSPORTED BY THE CIRCULATORY SYSTEM/RESPIRATION IS REGULATED BY THE BRAIN

1. Oxygen
2. Hemoglobin
3. Respiratory
4. Carbon dioxide
5. Phrenic
6. Hyperventilate
7. Blood pressure
8. CPR

SECTION XI. THE RESPIRATORY SYSTEM DEFENDS ITSELF AGAINST DIRTY AIR

1. Respiratory
2. Hair, mucous lining
3. Bronchial
4. Cilia
5. Macrophages
6. Lymph
7. Carbon
8. Disease
9. Lung

POST TEST

1. g
2. e
3. c
4. d

Copyright © 1992 W. B. SAUNDERS COMPANY All rights reserved

5. e
6. g
7. b
8. b
9. a
10. d
11. b
12. d

CHAPTER FIFTEEN

STUDY QUESTIONS

SECTION I. INTRODUCTION/THE DIGESTIVE SYSTEM CONSISTS OF THE DIGESTIVE TRACT AND ACCESSORY ORGANS

1. Nutrients
2. Metabolism
3. Alimentary
4. Mouth, anus
5. Gastrointestinal
6. Salivary, liver, pancreas

SECTION II. THE DIGESTIVE SYSTEM PROCESSES FOOD

1. Ingestion
2. Digestion
3. Absorption
4. Liver
5. Elimination

SECTION III. THE WALL OF THE DIGESTIVE TRACT HAS FOUR LAYERS

1. Digestive
2. Mucosa, epithelial
3. Digestion
4. Submucosa
5. Peristalsis
6. Connective
7. Parietal peritoneum
8. Peritoneal cavity
9. Peritonitis

SECTION IV. FOLDS OF THE PERITONEUM SUPPORT THE DIGESTIVE ORGANS

1. Mesentery
2. Intestine

Copyright © 1992 W. B. SAUNDERS COMPANY All rights reserved

3. Peritoneum
4. Greater omentum
5. Lesser omentum
6. Mesocolon

SECTION V. THE MOUTH INGESTS FOOD

1. Oral cavity
2. Tongue
3. Taste buds
4. Alveolar
5. Crown, roots
6. Dentin
7. Enamel
8. Pulp cavity
9. Parotid
10. Submandibular
11. Sublingual
12. Bolus

SECTION VI. THE PHARYNX IS IMPORTANT IN SWALLOWING

1. Swallowing
2. Pharynx
3. Oropharynx, nasopharynx, laryngopharynx
4. Soft palate
5. Tongue
6. Esophagus
7. Epiglottis

SECTION VII. THE ESOPHAGUS CONDUCTS FOOD TO THE STOMACH

1. Pharynx
2. Stomach
3. Peristaltic
4. Sphincter
5. Esophagus
6. Heartburn

SECTION VIII. THE STOMACH DIGESTS FOOD

1. Rugae
2. Contractions
3. Peristalsis
4. Mucus
5. Glands
6. Pepsin
7. Chyme
8. Pyloric
9. Small intestine

Copyright © 1992 W. B. SAUNDERS COMPANY All rights reserved

SECTION IX. MOST DIGESTION TAKES PLACE IN THE SMALL INTESTINE

1. Small intestine
2. Duodenum
3. Jejunum, ileum
4. Villi
5. Digestion
6. Duodenum
7. Pancreas
8. Absorbed

SECTION X. THE PANCREAS SECRETES ENZYMES

1. Pancreas
2. Exocrine
3. Pancreatic, digestive
4. Bile, duodenum

SECTION XI. THE LIVER SECRETES BILE

1. Liver
2. Liver cell
3. Lobe
4. Hepatic
5. Portal
6. Intestine
7. Gallbladder
8. Common bile duct

SECTION XII. DIGESTION OCCURS AS FOOD MOVES THROUGH THE DIGESTIVE TRACT

1. Chyme
2. Gastrin
3. Glucose
4. Mouth
5. Carbohydrate
6. Duodenum
7. Bile
8. Amino acids
9. Peptide
10. Stomach
11. Protein

SECTION XIII. THE INTESTINAL VILLI ABSORB NUTRIENTS

1. Villi
2. Villus
3. Lacteal
4. Absorbed
5. Liver
6. Fatty acids

Copyright © 1992 W. B. SAUNDERS COMPANY All rights reserved

SECTION XIV. THE LARGE INTESTINE ELIMINATES WASTES

1. Chyme
2. Ileocecal
3. Peristaltic
4. Cecum
5. Vermiform appendix
6. Appendicitis
7. Colon
8. Ascending colon
9. Transverse colon
10. Sigmoid colon, rectum
11. Anal canal

SECTION XV. A BALANCED DIET IS NECESSARY TO MAINTAIN HEALTH

1. Water
2. Minerals
3. Vitamins
4. Carbohydrates
5. Lipids
6. Proteins

POST TEST

1. b
2. d
3. e
4. f
5. a
6. c
7. b
8. a
9. e
10. a
11. d
12. c
13. e
14. c

Copyright © 1992 W. B. SAUNDERS COMPANY All rights reserved

CHAPTER SIXTEEN

STUDY QUESTIONS

SECTION I. INTRODUCTION

1. Waste products
2. Homeostasis
3. Excretion
4. Elimination
5. Circulatory
6. Urinary
7. Sweat glands
8. Urine

SECTION II. THE URINARY SYSTEM CONSISTS OF THE KIDNEYS, URINARY BLADDER, AND DUCTS

1. Kidneys
2. Bladder, ureters

Copyright © 1992 W. B. SAUNDERS COMPANY All rights reserved

3. Urine
4. Urethra

SECTION III. THE KIDNEYS CONSIST OF A CORTEX AND MEDULLA

1. Retroperitoneal
2. Kidney
3. Hilus
4. Renal capsule
5. Cortex, medulla
6. Nephrons
7. Urine
8. Blood pressure
9. Glomerular filtration
10. Sterile
11. Blood
12. Water
13. Diuretics

SECTION IV. URINE IS TRANSPORTED BY DUCTS AND STORED IN THE BLADDER

1. Ureters
2. Peristaltic contractions
3. Urinary bladder
4. Urethra
5. Prostate, penis
6. Vagina
7. Bladder

POST TEST

1. d
2. f
3. e
4. a
5. b
6. d
7. d
8. b
9. d
10. c
11. e

Copyright © 1992 W. B. SAUNDERS COMPANY All rights reserved

CHAPTER SEVENTEEN

STUDY QUESTIONS

SECTION I. INTRODUCTION

1. Homeostasis
2. Body fluids
3. Water
4. Electrolytes
5. Glucose, urea

SECTION II. THE BODY HAS TWO MAIN FLUID COMPARTMENTS

1. Water
2. Electrolytes
3. Intracellular
4. Extracellular
5. Blood pressure, osmotic concentration
6. Arteriole
7. Tissue fluid
8. Osmotic pressure
9. Lymphatic
10. Sodium ion
11. Potassium ion

SECTION III. FLUID INTAKE MUST EQUAL FLUID OUTPUT

1. Water
2. Blood
3. Catabolic
4. Kidneys
5. Digestive tract
6. Dehydration

SECTION IV. THE HYPOTHALAMUS REGULATES FLUID INTAKE

1. Hypothalamus
2. Osmotic
3. Thirst center
4. Fluids
5. Saliva

SECTION V. THE HYPOTHALAMUS REGULATES FLUID OUTPUT

1. Kidneys
2. ADH
3. Pituitary gland
4. Urine
5. Dehydrated
6. Plasma
7. Hypothalamus

Copyright © 1992 W. B. SAUNDERS COMPANY All rights reserved

SECTION VI. ELECTROLYTE BALANCE IS AFFECTED BY FLUID BALANCE

1. Electrolyte balance
2. Ions
3. Anions
4. Cations
5. Sodium
6. Water
7. Potassium

POST TEST

1. c
2. b
3. d
4. a
5. b
6. e
7. c
8. d
9. c
10. a
11. e
12. e

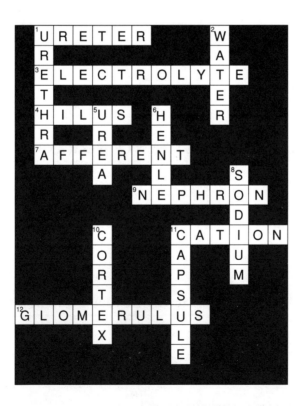

Copyright © 1992 W. B. SAUNDERS COMPANY All rights reserved

CHAPTER EIGHTEEN

STUDY QUESTIONS

SECTION I. THE MALE PRODUCES SPERM AND DELIVERS THEM INTO THE FEMALE

1. Sperm
2. Testes
3. Scrotum
4. Epididymis
5. Vas deferens
6. Spermatic cord
7. Semen
8. Prostate
9. Penis
10. Shaft, glans
11. Prepuce
12. Circumcision
13. Erect
14. Ejaculation
15. Testosterone

SECTION II. THE FEMALE PRODUCES OVA AND INCUBATES THE EMBRYO

1. Ovaries
2. Ova, estrogen, progesterone
3. Ovulation
4. Uterine
5. Uterus
6. Vagina
7. Cervix
8. Vulva
9. Mons pubis
10. Breasts
11. Lactation
12. Menstrual cycle
13. Menopause

SECTION III. FERTILIZATION IS THE FUSION OF SPERM AND OVUM

1. Uterus
2. Follicle
3. Ovum
4. Zygote
5. Ejaculation
6. Ovulation

SECTION IV. THE ZYGOTE GIVES RISE TO THE NEW INDIVIDUAL

1. Zygote
2. Embryo
3. Uterus
4. Fetal

Copyright © 1992 W. B. SAUNDERS COMPANY All rights reserved

5. Amnion
6. Placenta
7. Umbilical cord
8. Fetus

POST TEST

1. e
2. e
3. c
4. a
5. b
6. d
7. e
8. e
9. b
10. c
11. e
12. a
13. c
14. b
15. e

Copyright © 1992 W. B. SAUNDERS COMPANY All rights reserved